Regulating Next Generation Agri-Food Biotechnologies

Agri-food biotechnology policy and regulation is in transition from an early period focused on genetic engineering technologies to 'next generation' rules and regulatory processes. Can lessons learned from past and current regulatory oversights of agricultural biotechnology – and other high technology sectors – help us address new and emerging regulatory challenges in this sector?

The expert contributors in this volume discuss the experiences of a wide range of North American, European, and Asian countries with high technology regulation to address four key questions related to the past and future development of agri-food genomics regulation across the globe:

- How unique is agri-food biotecchology regulation, and how can it be evaluated using the existing tools of regulatory analysis developed in examinations of other sectors?
- Is a 'government to governance' model of regulatory regime development found in many other sectors relevant in this rapidly evolving sphere of activity?
- Is a stages model of regulatory regime development accurate? And, if so, at which stage are we currently positioned in the regulation of agri-food genomics products and technologies?
- What drives movement between stages in different countries and sectors? What are the key links between sectoral developments and more general macro- and micro-processes such as international relations and administrative behaviour?

By updating earlier empirical and theoretical social science perspectives on agricultural biotechnological regulation, this volume helps to inform future policy formulation. It will be of interest to practitioners and students of biotechnology, agriculture, science and technology policy, and regulatory processes more generally.

Michael Howlett is Burnaby Mountain Chair in the Department of Political Science at Simon Fraser University, and specialises in public policy analysis and resource and environmental policy. He has authored or edited more than twenty volumes, including most recently *Canadian Public Policy* (2012) and the *Routledge Handbook of Public Policy* (2012).

David Laycock is Professor in the Department of Political Science at Simon Fraser University, and focuses his research on political ideologies, democratic theory, Canadian party politics, public policy, and the politics of biotechnology. His published work includes *Policy Analysis in Canada* (2007) and *Representation and Democratic Theory* (2004).

Genetics and society

Series Editors: Ruth Chadwick, Director of Cesagen, Cardiff University; John Dupré, Director of Egenis, Exeter University; David Wield, Director of Innogen, Edinburgh University; and Steve Yearley, Director of the Genomics Forum, Edinburgh University.

The books in this series, all based on original research, explore the social, economic and ethical consequences of the new genetic sciences. The series is based in the Cesagen, one of the centres forming the ESRC's Genomics Network (EGN), the largest UK investment in social-science research on the implications of these innovations. With a mix of research monographs, edited collections, textbooks and a major new handbook, the series is a valuable contribution to the social analysis of developing and emergent bio-technologies.

Series titles include:

New Genetics, New Social Formations
Peter Glasner, Paul Atkinson and Helen Greenslade

New Genetics, New Identities
Paul Atkinson, Peter Glasner and Helen Greenslade

The GM Debate
Risk, politics and public engagement
Tom Horlick-Jones, John Walls, Gene Rowe, Nick Pidgeon, Wouter Poortinga, Graham Murdock and Tim O'Riordan

Growth Cultures
Life sciences and economic development
Philip Cooke

Human Cloning in the Media
Embryonic stem cell research in India
Aditya Bharadwaj and Peter Glasner

Handbook of Genetics and Society
Paul Atkinson, Peter Glasner and Margaret Lock

The Human Genome
Chamundeeswari Kuppuswamy

Community Genetics and Genetic Alliances
Eugenics, carrier testing and networks of risk
Aviad E. Raz

**Neurogenetic Diagnoses. The Power of Hope and the
Limits of Today's Medicine**
Carole Browner and H. Mabel Preloran

Debating Human Genetics
Contemporary issues in public policy and ethics
Alexandra Plows

Genetically Modified Crops on Trial
Opening up alternative futures of Euro-agriculture
Les Levidow

Creating Conditions
The making and remaking of a genetic condition
Katie Featherstone and Paul Atkinson

Genetic Testing
Accounts of autonomy, responsiblility and blame
Michael Arribas-Allyon, Srikant Sarangi and Angus Clarke

Regulating Next Generation Agri-Food Biotechnologies
Lessons from European, North American, and Asian experiences
Edited by Michael Howlett and David Laycock

Forthcoming titles include:

Scientific, Clinical and Commercial Development of the Stem Cell
From radiobiology to regenerative medicine
Alison Kraft

Barcoding Nature
Claire Waterton, Rebecca Ellis and Brian Wynne

Gender and Genetics
Towards a sociological account of prenatal screening
Kate Reed

Regenerating Bodies
Tissue and cell therapy in the twenty-first century
Julie Kent

Regulating Next Generation Agri-Food Biotechnologies

Lessons from European, North American, and Asian experiences

Edited by Michael Howlett and David Laycock

Routledge
Taylor & Francis Group

LONDON AND NEW YORK

FONDATION
BROCHER

First published 2012
by Routledge

2 Park Square, Milton Park, Abingdon, Oxfordshire OX14 4RN

Simultaneously published in the USA and Canada
by Routledge

711 Third Avenue, New York, NY 10017

First issued in paperback 2014

Routledge is an imprint of the Taylor and Francis Group, an informa business

British Library Cataloguing in Publication Data
A catalogue record for this book is available from the British Library

Library of Congress Cataloging-in-Publication Data
Regulating next generation agri-food bio-technologies: lessons from European,
North American and Asian experiences/edited by Michael Howlett and
David Laycock.
 p. cm. – (Genetics and society)
 Includes bibliographical references and index.
 1. Agricultural biotechnology–Government policy. 2. Agriculture and state.
 3. Agricultural biotechnology–Europe. 4. Agricultural biotechnology–
 North America. 5. Agricultural biotechnology–Asia. I. Howlett, Michael,
 1955– II. Laycock, David H., 1954–
 S494.5.B563R46 2012
 630–dc23
 2011040703

ISBN 978-0-415-69361-5 (hbk)
ISBN 978-1-138-02009-2 (pbk)
ISBN 978-0-203-12332-4 (ebk)

Typeset in Times New Roman
by Sunrise Setting Ltd

Contents

List of figures   xii
List of tables   xiii
Contributors   xiv
Acknowledgements   xx

Introduction   1

**1   Regulating next generation biotechnologies:
tentative regulation for emerging technologies**   3
DAVID LAYCOCK AND MICHAEL HOWLETT

**PART 1
First and second generation agri-food genetic
technologies and regulatory regimes: issues
and overviews**   13

**2   Generating regulatory futures: from
agbiotech blockages to a bioeconomy?**   15
LES LEVIDOW

Introduction: beyond biotech blockages? 15
Co-production of technology with socio-natural order 15
Making Europe safe for agbiotech 17
Putting agbiotech on trial, reshaping regulations 20
Designing next generation biotech for a bioeconomy 26
Conclusions: lessons for future technology? 29

3 **Learning from experience: how do we use what
we have learned to reform regulatory oversight
of new agricultural biotechnologies?** 34

ALAN MCHUGHEN

Introduction 35
International trade implications 38
Scientific flaws in agbiotech regulatory scrutiny 38
The purpose of agbiotech 'science based' regulations 39
Placing the risks of agbiotech in context. How does
the scientific community evaluate the hazards of
rDNA crop and plant breeding? 41
How do we properly and safely manage products of
agricultural biotechnology? 43
Conclusions 44

PART 2
Regulatory regime development theory and practice 47

4 **Regulatory lifecycles and comparative
biotechnology regulation: analysing regulatory
regimes in space and time** 49

MICHAEL HOWLETT AND ANDREA MIGONE

Introduction: understanding the spatial and
temporal dimensions of national regulatory regimes 49
Regulation, regulatory regimes and regulatory lifecycles 50
The agri-food biotechnology/genomics case 56
Understanding the evolution of biotechnology
regulation: the temporal dimension 60
Conclusion 64

5 **Pragmatism revisited: an overview of the
development of regulatory regimes of
GMOs in the European Union** 73

ANDERS JOHANSSON

Introduction 73
The regulatory regimes in European GMO politics 75
Policy models in European GMO politics 84
Mapping the regimes of European GMO politics 87

PART 3
GMO regulatory regimes in practice:
Europe, Asia, and North America 93

6 **Contested frames: comparing EU versus**
US GMO policy 95
SARAH LIEBERMAN AND ANTHONY R. ZITO

Introduction 95
The argument 97
USA policy history 101
The WTO arena: biotech products 104

7 **The global battle over the governance of agricultural**
biotechnology: the roles of Japan, Korea, and China 111
YVES TIBERGHIEN

Introduction 111
Theoretical framework 112
Japan: political reversal and regulatory adaptation 115
China: representation without institutions 119
Conclusion 125

8 **The EU's governance of plant biotechnology**
risk regulation: still contested, still distinct 128
PAULETTE KURZER AND GRACE SKOGSTAD

The EU GMO risk regulation framework 130
EU regulatory debates and continuing contestation 134
Breaking the log jam through regulatory change? 137
Conclusion 141

PART 4
Lessons from other high technology sectors 147

9 **Regulating nanotechnology in China:**
governance, risk management, and
regulatory effectiveness 149
DARRYL S. L. JARVIS AND NOAH RICHMOND

Introduction 149
Science and nanotechnology in China 150
Managing nano-based risks in China: regulatory responses 153
Nano safety regulation in China: EHS risk 156
Regulatory effectiveness and governance gaps 157

10 **Lessons from biomedical technology regulation:**
    North American and European comparisons   164

   ISABELLE ENGELI, CHRISTINE ROTHMAYR ALLISON
   AND FRÉDÉRIC VARONE

   Introduction 164
   Theorising policy trajectories: path dependency,
      policy transfer, and scientific development 165
   Comparing regulatory trajectories for
      ART and embryo related research 166
   Case studies: explaining trajectories 170
   Conclusion 177

   PART 5
   Agricultural biotechnologies and the public:
   deliberation, opinion, ethics, and participation   183

11 **Network deliberations, advocacy groups,**
    and the legitimacy of the European Union   185
   ÉRIC MONTPETIT

   Introduction 185
   Democratic norms 186
   EU institutions and democratic norms 188
   The biotechnology actor survey 189
   Elections 190
   Horizontal networks versus hierarchical settings 193
   Deliberation 194
   Inclusiveness of perspectives 196
   Conclusion 200

12 **Getting to maybe: assessments of benefits and**
    risks in Canadian public opinion on
    biotechnological innovation   203
   STEVEN WELDON, DAVID LAYCOCK, ANDREA NÜSSER
   AND COLIN WHELAN

   Introduction 203
   Charting support for biotechnologies 204
   Past research on sources of support for biotechnology 206
   A unified causal model of opinion formation 208
   Data and methods 211

Modelling perceptions of benefits and risks 212
Modelling support for biotechnology by risk/benefit group 214
Discussion and conclusion 216

**13 Deriving policy and governance from
deliberative events and mini-publics**   220

MICHAEL M. BURGESS

Introduction 220
Research on deliberative engagement using mini-publics 222
Discussion of analyses across deliberations:
   governing uncertainty 230
Conclusion 233

**14 Second generation governance for second
generation GM**   237

CHRISTOPH REHMANN-SUTTER

Discordance 238
The producer–consumer relationship 243
Four concepts of governance 245
The role of (bio)ethics in governance 249
Conclusions: ethical governance 252

Index   256

List of figures

3.1 Chart comparing the probability of different forms of plant breeding giving rise to unexpected effects (as a proxy for potentially hazardous outcomes) 42

9.1 Policy and key regulatory agencies for nanotechnology in China 153

9.2 Risk regulation of nanoscience and nanotechnology in China 154

10.1 Patterns in regulatory trajectory over ART 167

10.2 Patterns in regulatory trajectory over embryo-related research 168

11.1 Comparison of the most important factors conferring legitimacy to policy decisions 191

11.2 Most important legitimizing factor for actors involved in the EU and actors involved in nation states exclusively 192

11.3 Agreement with the suggestion that decisions are made in networks rather than in hierarchical settings 194

11.4 Agreement with the suggestion that policymakers can be persuaded to change their mind 195

11.5 The extent to which advocacy groups agree that policy decisions are preceded by deliberation with a wide range of actors 197

11.6 Frequency whereby industry, experts and government employees meet with people holding different views 198

11.7 Ratio estimations of precautionary/promotional attitudes toward biotechnology 199

12.1 Modelling support for biotech 209

List of tables

4.1	The linear model after Otway and Ravetz (1984)	52
4.2	Stages of risk controversy after Leiss (2001)	52
4.3	Stages of regulatory development after Bernstein (1955), Otway and Ravetz (1984) and Leiss (2001)	53
4.4	Early stage regulatory lifecycle model	54
4.5	Nine-stage model of regulatory regime lifecycle	55
4.6	Paarlberg model of policy options and regimes towards GM crops	58
4.7	Mapping comparative biotechnology regulatory regimes	59
4.8	The GM biotechnology lifecycle in the US and EU	63
4.9	The regulatory issue field in biotechnology	65
5.1	Key events in EU biotech regulation	74
5.2	Regimes of European GMO politics	89
9.1	List of nanotechnology related standards in China	160
12.1	Canadians' approval of different biotechnology applications	205
12.2	Typology of orientations toward biotechnologies in Canada	212
12.3	Linear regression models of benefits and risks	213
12.4	Predicted support for biotechnologies by risk-benefit groupings	215

Contributors

Editors

Michael Howlett, Professor and Burnaby Mountain Chair, Department of Political Science, Simon Fraser University. Dr Howlett specialises in public policy analysis, political economy, and resource and environmental policy. He taught at Queen's University (1986–88) and the University of Victoria (1988–89) before coming to SFU, and was Visiting Professor at the Lee Kuan Yew School of Public Policy of the National University of Singapore (2009–10). He is the author of *The Political Economy of Canada* (1992 and 1999), *Studying Public Policy* (1995, 2003 and 2009),*Canadian Natural Resource and Environmental Policy* (1997 and 2005), *In Search of Sustainability* (2001), *Integrated Policy Making for Sustainable Development* (2009), *Designing Public Policy* (2010), and co-author of *The Public Policy Primer* (2010). He has published numerous articles and chapters on public policy analysis, political economy, Canadian politics and resource and environmental policy, and he currently sits on several journal editorial boards.

David Laycock, Professor, Department of Political Science, Simon Fraser University. Dr Laycock focuses his research on political ideologies, democratic theory, Canadian party politics, public policy, and the politics of biotechnology. His published work includes *Populism and Democratic Thought in the Canadian Prairies* (1990), *The New Right and Democracy in Canada* (2001), *Representation and Democratic Theory* (2004), and articles or chapters on direct democracy, various Canadian political parties, populism, representation and public opinion on biotechnology innovation. With Michael Howlett and Laurent Dobuzinskis, he edited *Policy Studies in Canada* (1996) and *Policy Analysis in Canada* (2007).

Chapter authors

Mike Burgess, Professor in Biomedical Ethics, W. Maurice Young Centre for Applied Ethics and Department of Medical Genetics, University of British Columbia. Dr Burgess combines ethical and social theory with social science methods related to policy in health care, health research and biotechnology. His research has been funded by SSHRC, CIHR, Genome Canada and Genome BC, as well as NGOs and industry contracts. Most recently, he has focused on science and technology policy and public engagement based on theories of deliberative democracy. With Peter Danielson he is Principal Investigator of the Genome Canada/Genome BC-funded "Building a GE3LS Architecture" and is a collaborator on other genome science projects in microbial genomics of forest soils (PI: W. Mohn) and military explosives (PI: L Eltis). Dr Burgess worked with Kieran O'Doherty to implement deliberative engagements on biobanks in BC, at the Mayo Clinic in Minnesota (PI: B. Koenig) and in Western Australia.

Isabelle Engeli, Lecturer, Department of Political Science, University of Geneva. Dr Engeli's main areas of expertise are in comparative public policies, agenda setting and policy change, as well as political representation and gender. She is a member of the Comparative Policy Agendas Project, which aims at developing systematic quantitative analysis of issue attention and politicisation in Europe and North America across time. She also co-leads a systematic comparative study of politicisation and policy making on morality issues such as euthanasia, embryo and stem cell research and same-sex marriage across West European countries. In 2008–10, she was Max Weber fellow at the European University Institute. Her recent publications include *Les Politiques de la Reproduction* (2010); "Policy Struggle on Reproduction: Doctors, Women, and Christians", *Political Research Quarterly*; "The Challenges of Abortion and Assisted Reproductive Technologies Policies in Europe", Comparative European Politics, and "Gender Gap and Turnout", *Swiss Review of Political Science.*

Darryl Jarvis, Associate Professor, Lee Kuan Yew School of Public Policy, National University of Singapore. Dr Jarvis specialises in risk analysis and the study of political and economic risk in Asia, including investment, regulatory and institutional risk analysis. He is author and/or editor of several books and has contributed articles to leading international journals. He has been a consultant to various government bodies and business organisations, and for two years was a member of the investigating team and then chief researcher on the Building Institutional Capacity in Asia (BICA) project commissioned by the Ministry of Finance, Japan. His current research is a large cross-national study of risk causality in four of Asia's most dynamic industry sectors. He teaches courses on risk analysis, markets and international governance, and international political economy.

Anders Johansson, Department of Water and Environmental Studies, Linköpings universitet, Sweden. In 2010–11 Dr Johansson was a post-doctoral fellow at Simon Fraser University working on topics related to public participation and biotechnology policy making. In 2009, he completed his PhD dissertation, *Biopolitics and Reflexivity: A Study of GMO Policy making in the European Union,* at Linköpings universitet.

Paulette Kurzer, Professor, Department of Political Science, University of Arizona. Dr Kurzer teaches courses in comparative politics, European politics and advanced industrialised states. Her research interests include institutions of the European Union and the intersection between domestic and European policy making. Her work has appeared in the *Journal for European Public Policy, Journal of Common Market Studies*, and other refereed journals, as well as several edited volumes. Her most recent book is *Markets and Moral Regulation* (2001). She is currently working on the perception of risk and the formation of a new public health domain in the European Union.

Les Levidow, Research Fellow, The Open University, United Kingdom. Dr Levidow has been managing editor of *Science as Culture* since its inception in 1987, and its editor since 2005. His research interests include controversial agricultural technologies, especially agbiotech and biofuel crops, as well as quality alternatives to agri-industrial systems. These topics provide case studies for several policy-relevant issues: agri-environmental sustainability, European integration, trade conflicts, governance, public participation, regulatory science and the precautionary principle. His most recent books are *GM Food on Trial: Testing European Democracy* (2010) and *Governing the transatlantic conflict over agricultural biotechnology: contending coalitions, trade liberalisation and standard setting* (2006, with Joseph Murphy).

Sarah Lieberman, Senior Lecturer in Politics, Faculty of Applied Social Sciences, Canterbury Christchurch University, United Kingdom. Dr Lieberman has published articles on the EU's licensing of new genetically modified (GM) products by the European Union; on the conflict between the USA and the EU over genetically modified organisms (GMO); on the American WTO challenge to the EU's moratorium on biotech products; on the impact on developing countries of the USA/EU conflict over GMOs; on participatory decision making on GMOs in France and Britain; and for Greenpeace on trade in genetically engineered commodities.

Alan McHughen, Professor, Department of Botany and Plant Sciences at the University of California, Riverside. Alan McHughen is a public sector educator, scientist and consumer advocate. A molecular geneticist with an interest in crop improvement and environmental sustainability, he helped develop US and Canadian regulations covering genetically engineered crops and foods. He has served on recent US National Academy of Sciences panels investigating the environmental effects of transgenic plants, and investigating the health effects of genetically modified foods. He is now Past President and Treasurer of the International Society for Biosafety Research (ISBR). Having developed internationally approved commercial crop varieties using both conventional breeding and genetic engineering techniques, he has firsthand experience with the relevant issues from both sides of the regulatory process. His *Pandora's Picnic Basket; The Potential and Hazards of Genetically Modified Foods* (2007) addresses the myths and genuine risks of genetic modification (GM) technology.

Andrea Migone, Lecturer, Department of Political Science, Simon Fraser University. Dr Migone's areas of interest are public administration and comparative political economy. He has served as a Post-Doctoral fellow and worked on a number of projects in the biotechnology funded by GenomeBC and GenomeCanada and his work on regulation ad participation in biotechnology has appeared in *Science and Public Policy, Technology in Society*, and *Policy and Society*.

Éric Montpetit, Professor and Chair, Département de science politique, Université de Montréal. Dr Montpetit completed a PhD. in comparative and Canadian public policy at McMaster University in 1999. His current research concerns the capacity of policy actors to agree on policy choices despite differences in opinion; the use of science in policy making; and biotechnology policy. Past research has included environmental policy comparisons. In 2006, he received the American Political Science Association's Lynton Keith Caldwell Prize for the best book on environmental politics and policy for his book *Misplaced Distrust*. He has published his work in major journals including the *Journal of European Public Policy*, *Administration & Society, Comparative Political Studies* and *Governance*.

Andrea Nüsser, PhD candidate, Department of Political Science, University of British Columbia. Ms Nüsser completed her BA in Political Science from the University of Bonn and her MA in the Department of Political Science at Simon Fraser University. Her research is on public opinion and comparative party politics, on which she has published an article with Steven Weldon in *German Politics and Society*. In 2009–10 she was a research assistant for David Laycock and Steven Weldon on projects dealing with Canadian public opinion with regard to genomic innovation.

Christoph Rehmann-Sutter, Head, Unit for Ethics in the Biosciences, University of Lübeck. Dr Rehmann-Sutter is Professor of Theory and Ethics in the Biosciences, University of Lübeck, Titular Professor of Philosophy at the University of Basel and Visiting Professor at the London School of Economics. He has trained in molecular biology, subsequently in philosophy and sociology, and has widely published in different fields of bioethics. From 2001–09 he was President of the Swiss National Advisory Commission on Biomedical Ethics. His current research interests and publications include the philosophical foundations of bioethics, end-of-life issues in palliative care, the ethical implications of genomics, regenerative medicine and synthetic biology. In his research, he combines qualitative empirical approaches to participants' views with theoretical normative work in ethics. His published work includes co-edited volumes *Genes in Development: Re-reading the molecular paradigm* (2006); *Bioethics in Cultural Contexts. Interdisciplinary Approaches* (2006); *The Contingent Nature of Life: Bioethics and the Limits of Human Existence* (2008); and the research monograph *Zwischen den Molekülen* (2005).

Noah Richmond, Research Associate, Lee Kuan Yew School of Public Policy, National University of Singapore.

Christine Rothmayr Allison, Associate Professor of Political Science, University of Montreal. Dr Rothmayr Allison's two main fields of interest are comparative public policy, focusing on the fields of biotechnology, biomedicine and higher education, and courts and politics, in particular, the impact of court decisions on public policy making in North America and Europe. She has published articles in the *European Journal of Political Research, Comparative Political Studies* and *West European Politics*. With Éric Montpetit and Frédéric Varone, she edited *The Politics of Biotechnology in North America and Europe: Policy Networks, Institutions and Internationalization* (2007). Before joining the University of Montreal, she taught at the University of Geneva, Switzerland. She received her PhD from the University of Zurich, Switzerland.

Grace Skogstad, Professor of Political Science, Department of Political Science, University of Toronto (St. George) and the Department of Social Sciences (UTSC). Dr Skogstad's areas of interest include comparative public policy; the role of international institutions in regulating transatlantic agricultural and food trade disputes; Canadian federalism and intergovernmental relations; policy networks and governance; the role of ideas in policy making; and the impact of economic globalisation and political internationalisation on the domestic politics and governance of agriculture and food. She is the author of *The Politics of Canadian Agricultural Policy* (1987) and *Internationalization and Canadian Agriculture: Policy and Governing Paradigms* (2008) as well as numerous book chapters and articles. She is also the co-editor of *Canadian Federalism: Performance, Effectiveness and Legitimacy* (2002, 2007), and *Policy Communities and Public Policy* (1990).

Yves Tiberghien, Associate Professor in the Department of Political Science, University of British Columbia. Dr Tiberghien specialises in comparative political economy (Japan, Korea, China, and the European Union) as well as in international relations (global governance, globalisation). Much of his work centres on the tensions between the pursuit of prosperity (the market) and the pursuit of democratic legitimacy. His most recent research focuses on the role of China in the shifting global power and governance in the next decades, and on the effects of the globalisation on widening economic inequality in Japan. He the author of *Entrepreneurial States: Reforming Corporate Governance in France, Japan, and Korea* (2007), and several forthcoming books: *L'Asie, le G20, et le future du Monde; The Global Battle over the Governance of Genetically-Engineered Food*, and *Minerva's Rule: Canadian, European, and Japanese Leadership in Global Institution-Building*, edited with Julian Dierkes.

Frédéric Varone, Professor, Political Science, University of Geneva. Dr Varone was a lecturer at the Institute of Political Science at the University of Bern from 1991 to 1995 where he carried out several evaluation studies in the area of energy and technology policy. He taught at the Institute for Advanced Studies in Public Administration (1997–99), at the University of Lausanne from 1998, then at the Catholic University of Louvain (Louvain-la-Neuve, Belgium) until 2006. Since 2006, he has been Professor of Political Science at the University of Geneva (Switzerland). He teaches public administration, comparative policy analysis, programme evaluation and public management at Bachelor,

Master and PhD levels. He is currently involved in various Executive Masters of Public Management (e.g. Ecole Solvay at the Free University of Brussels, Business School of the University of Salzburg, and Centre of Competence for Public Management of the University of Bern). Among his many publications are *Public Policy Analysis* (2007), with Peter Knoepfel, Corinne Larrue and Michael Hill; *Régulation publique des biotechnologies: Biomédecine et OGM agroalimentaires en Belgique et en France* (2005), with Nathalie Schiffino; and *Le choix des instruments des politiques publiques. Une analyse comparée des politiques d'efficience énergétique du Canada, du Danemark, des Etats-Unis, de la Suède et de la Suisse* (1998).

Steven Weldon, Assistant Professor, Department of Political Science, Simon Fraser University. Dr Weldon specialises in comparative politics. His research focuses on political parties and party systems, public opinion, political behaviour, and comparative ethnic relations and conflict. Currently he is engaged in two main research projects. The first examines how intra-party politics and patterns of party member activism shape the larger democratic process. The second focuses on the formation of political community in ethnically divided societies, including the integration of immigrants and ethnic minorities in advanced democracies. His work has appeared in such journals as the *American Journal of Political Science*, the *British Journal of Political Science, Party Politics, West European Politics*, and *Rivista italiana di scienza politica*.

Colin Whelan, BA, Political Science, Simon Fraser University. Mr Whelan completed his BA in Political Science and Philosophy in June 2011, and was awarded the Dean's medal for highest academic achievement in the Faculty of Arts and Social Sciences. In 2010–11 he was a research assistant for David Laycock and Steven Weldon on public opinion with regard to genomic innovation.

Anthony R. Zito, Reader of Politics, Newcastle University. Dr Zito's research interests include the European Union, domestic environmental politics and policy making, European political economy, European public policy, international environmental politics and policy making, international political economy, and international organisations. His current 'Comparing Environmental Agencies and Governance' project compares the evolution of environmental agencies in the new Millennium. He has published numerous book chapters, and articles in the *Journal of European Public Policy, Political Studies, Politische Vierteljahresschrift, Public Administration, Environmental Politics,* and *Globalizations*. He is currently writing a monograph that summarises the complete empirical results of the ESRC Future Governance Programme Project with Dr Rüdiger Wurzel and Dr Andrew Jordan.

Acknowledgements

A joint endeavour such as this relies on contributions from many people. Our first debt of gratitude is to Dr Sarah Hartley, who proposed back in April 2009 that we should submit an application to the Brocher Foundation competition for a conference in Hermance, Switzerland. Sarah helped to draft our successful proposal, and secured the support of Genome BC for travel funding. Dr Yves Tiberghien also contributed important advice to our proposal, drawing on his many contacts in Europe and Asia involved in the regulation and politics of biotechnologies.

The Brocher Foundation[1] hosted our two-day symposium on emerging agricultural biotechnology governance challenges in July 2011 at their picturesque Hermance facility, and generously arranged and financed participants' hotel accommodations and meals. As our volume would not have come together without this symposium, we are extremely grateful to the Brocher Foundation for its support, particularly Marie Grosclaude. Travel costs for participants were paid for by a combination of Genome BC, Genome Canada, the Lee Kuan Yew Public Policy School at the National University of Singapore, Simon Fraser University, and the University of British Columbia. We thank them all.

For their excellent original papers and contributions to lively discussions, we thank all those at the symposium. Participants included not only the authors of the chapters in this volume, but also Catherine Lyall, ESRC Innogen Centre, University of Edinburgh; Hideaki Shiroyama and Makiko Matsuo, Graduate School of Public Policy, Tokyo University; Jorgen Schlundt, Director, Department of Food Safety, Zoonoses and Foodborne Diseases, World Health Organization, Geneva; Laurence Boisson de Chazournes, Director, Department of Public International Law and International Organization, Faculty of Law, University of Geneva; Francois Pythoud, Head, International Sustainable Agriculture Unit, Swiss Federal Office of Agriculture; Anne-Gabrielle Wust Saucy, Head of Biotechnology Section, Swiss Federal Office for the Environment; and Robson de M. Fernandes, World Trade Organization.

On the publication process side, our first debt is to Daisy Laforce. Her formidable editing and precise analytic skills enabled our contributors to make important revisions crucial to production of a strong, clear, and coherent collection. In addition, as our genomics social science project administrator, Daisy helped to organise the Brocher symposium, and managed post-conference administrative work.

SFU post-doctoral fellow, Dr Anders Johansson, also made important contributions to the volume's thematic coherence. Brian Wynne (Lancaster University) and Helen Greenslade (Cardiff University) with the ESRC Centre for the Economic and Social Aspects of Genomics (CESAGen) supported our proposal to publish this volume in Routledge's Genetics and Society series.

Finally, we wish to thank Routledge publishers for this opportunity to share this research with a wider audience.

Michael Howlett and David Laycock
Simon Fraser University
June 2011

Note

1 The Brocher Foundation's mission is to encourage research on the ethical, legal, and social implications of new medical technologies. Its main activities are to host visiting researchers and to organise symposia, workshops, and summer schools. More information on the Brocher Foundation programme is available at www.brocher.ch.

Introduction

1 Regulating next generation biotechnologies

Tentative regulation for emerging technologies

David Laycock and Michael Howlett

This book is about learning from experience. The question animating all of the following contributions is straightforward but demanding. How do lessons learned from past and current regulatory oversight of agricultural biotechnology and other high technology sectors help us – academics, policy practitioners, political decision makers, and the public – address new regulatory challenges in the agri-food genetics sector?

The book begins from the observation that agri-food biotechnology policy and regulation is undergoing a transition from early periods focused on genetic engineering technologies, especially the development and commercialisation of genetically modified organisms (GMOs), to 'next generation' rules and processes linked to innovations in 'omics technologies (genomics, proteomics, metabolomics, and transcriptomics). These innovations are resulting in diverse new applications in everything from crop-based pharmaceuticals to a wide range of industrial products. The accompanying regulatory transition is occurring in a range of fields, including 'red' (bio-medical) genomics, 'white' (industrial) uses, 'black' (or resource-oriented applications) and the more familiar 'green' or agri-food uses (Bauer 2005; Migone and Howlett 2009; Howlett and Migone 2010a). As a whole, our volume argues that next generation developments in policy and regulation should be informed by a clear understanding of the underlying technologies and technological changes that largely drive them forward. It is also crucial, however, that emerging regulations rely on careful comparative historical analyses of the impact and results of earlier moves in this area. Comparative, cross-sectoral study is useful for these purposes, as is the study of activities in related technology-driven fields such as nano-technology and/or assisted reproductive technologies. Both kinds of study can make key contributions to the development of a better understanding of regulation and regulatory processes in rapidly evolving spheres of scientific/industrial endeavour and are a primary feature of the essays in this volume. On such bases, the socio-cultural as well as the technological foundations of biotech innovations can be better appreciated and assessed.

Our contributors focus on regulatory developments within the agri-food sector in the European Union, North America, and Asia. This sector was the site of much controversy in earlier periods surrounding the development and deployment of genetically modified organisms, but has to some degree moved beyond GMOs to a range of other techniques and processes expected to enhance agricultural production and food processing.

The deployment of genomic-based biotechnologies in the agri-food sectors in many countries beginning in the 1980s presented huge regulatory and other challenges around the world. After the introduction of genetically-engineered food revealed legitimacy and other gaps with traditional regulatory systems (Levidow and Carr 2010; Lyall 2007; Lyall *et al.* 2009), development of 'second generation' technologies, such as the use of biomarkers for selective breeding, raised additional questions about effective monitoring and control of 'omic applications (Mintrom and Williams 2009). These concerns have been exacerbated by the development of 'third' generation technologies in areas such as animal cloning and synthetic biology in recent years. The authors in the volume all address, in one way or another, the question of how societies, governments, and evolving international regulatory regimes can deal with these new and near future technologies.

The chapters in this book explore governance challenges associated with biotechnological advances. The technologies and regulations associated with the first generation, which focus on products such as genetically modified crops and platforms such as the human genome project, are still very relevant to agricultural economies throughout the world. However, much is changing at a striking pace. In the agricultural sector, for example, second and third generation applications, such as genomic biomarkers for diagnostic and selective breeding purposes, animal cloning and genetic modification, commercial development of biofuels, the use of nano-technologies in plant and animal production, and pharmaceutical crops are all challenging existing regulatory regimes and the societies in which they are politically embedded (Howlett and Migone 2010a; Howlett and Migone 2010b). But while some countries, especially in the developing world, clamour for access to the patents and other property right determinants from the dissemination of these technologies and products, others discourage their use, with potentially large costs to their scientific and economic development (Sleeboom 2001; Sheingate 2006; Potrykus 2010). Why these differences exist and how they are being managed is a major theme of this collection.

From a governance perspective, the development and deployment of second and third generation biotechnologies raise many key questions for agri-food regulators. Experience with the governance of genetically modified (GM) crops offers some lessons about the importance of linkages between public opinion and regulation for subsequent applications, for example, but has left many questions unanswered about appropriate governance regimes for the new biotechnologies. These include both practical questions related to monitoring and enforcement, such as traceability regulations, the tension between science-based international trade commitments and democratic values, as well as a host of other concerns that challenge the legitimacy of both the old, and any new, regulations (Skogstad 2003; Sjoberg 2005). While transgenic technology has been widely adopted in North America, first wave agri-food applications became the target of a global resistance movement including non-governmental organisations (NGOs), key states (the European Union (EU), Japan, Korea), and international organisations

(Mwale 2006; Legge and Durant 2010). Biotechnological innovation has been criticised on grounds as varied as innattention to food safety, insufficient accountability to citizens via product labelling, threats to biodiversity and the environment, placing scientific progress ahead of the public interest, and enhancing the power of large global corporations vis-à-vis poorer countries and consumers. Governance regimes for next generation biotechnologies will continue to encounter legitimation difficulties unless they address the political, cultural, technical, coordination, and practical issues that have accompanied earlier waves of biotechnological innovations (Ayele 2007; Jasanoff 2005).

The qualitatively different environmental, ethical, legal, and social implications entailed by applications of new technologies require elaboration, investigation, and analysis. Lessons can be learned, for example, from comparative examination of notable successes and failures in regulatory and other responses to earlier generations of technologies. Relevant evidence can be drawn from experiences in the USA, Canada, the countries of the EU, as well as industrialised or rapidly industrialising countries in Asia and Latin America. The comparative lessons drawn by the authors in this volume concern regulatory design and application, and how public opinion research and public engagement processes can identify foundations of more legitimate governance frameworks and practices in relation to new biotechnologies.

As many of the essays in the volume suggest, responding to the unique characteristics and potential effects of new biotechnologies may call for new modes of genomic governance more suitable to the smaller scale, more diverse, and more widespread application of genomic knowledge that characterise recent developments (Bingham 2008; Montpetit and Rouillard 2008). Designing appropriate governance modes for new biotechnologies requires us to learn much more about the fate of such processes in earlier periods of regulatory activity. This volume addresses gaps in our understanding of earlier periods of 'omics regulation, covering the role played by public opinion, characteristic patterns and drivers of regulatory regime formation and development, and cross-national and cross-sectoral variations in those same regimes.

The book has its origins in several multi-year projects funded in Canada by Genome BC and Genome Canada in which the two editors served as principal investigators dealing with social science components and aspects of agri-food second and third wave genomics innovations. A symposium held in Switzerland under the auspices of the Geneva-based Brocher Foundation in July 2010 brought together scholars and industry, NGO and government practitioners from North America, Europe, Japan, Korea and Singapore to discuss past practices and potential future developments in agri-food genomics technologies and regulations. Symposium panellists and audience participants included 20–25 legal and government experts from several European countries and Geneva-based international organisations.

The symposium featured a productively focused, well-rounded assessment of the governance achievements and challenges in this field and provided an overview of key aspects of the emerging governance landscape for new genomics technologies. Participants argued that these would be informed by the legacies of past practices in the area, and that such developments would build on established rules and regulations that have so far helped prevent outcomes predicted by the direst forecasts concerning employment of genomics techniques inside and outside the lab. Participants anticipated as well that existing regimes that stress the use of the precautionary principle would often not be able to deal effectively with less worrisome or less intrusive technologies, especially those widely perceived as positively beneficial in the bio-medical area.

The following chapters are based on papers presented at the Brocher symposium. They review best practices and problem cases, identify crucial areas in which additional research and public discussion are required, and examine possible ways of enhancing the legitimacy of existing rules and regulations via novel participatory governance techniques. Dealing with a wide range of North American, European and Asian countries, our expert contributors together construct a wide multidisciplinary lens to address four key questions related to the past and future development of agri-food genomics regulation across the globe. These include:

- Is agri-food biotechnology regulation *sui generis*? Is it unique and idiosyncratic or can it be evaluated using the existing tools of regulatory analysis developed in examinations of other sectors?
- Is a 'government to governance' model of regulatory regime development, in which trends and tendencies towards more collaboration and less hierarchical state-societal relations are a dominant motif, relevant to this rapidly evolving sector?
- If a stages model of regulatory regime development is accurate, then at which stage are we in the regulation of agri-food genomics products and technologies?
- What drives changes between stages in different countries? And in assessing drivers, what are the key links between sectoral (meso) developments and more general macro- (and micro-) developments such as international relations and administrative behaviour?

With respect to the first question, the authors come down firmly in the negative. That is, rather than being viewed as a unique and idiosyncratic set of regulatory practices, the essays in this volume argue the opposite: that a general, discernible pattern of regulatory development exists and applies to this sector. Thus, much apparent variation in national practices stems from the specific stage of regulatory regime development that characterises a country's circumstances, rather than from fundamental differences between competing regulatory regime types.

Concerning the second question, contributors argue that much recent thinking on the evolution of most regulatory regimes does tend towards the promotion of increased public participation and collaboration and away from more traditional state-directed regime elements and that such trends and tendencies are also apparent, and useful, in the area of next generation genomics regulation. No one believes that governments' substantial past roles in the promotion and development of scientific research and industrial and commercial applications of genomics technologies are ending. However, our authors acknowledge, in various ways, that increasing concerns for legitimacy over the past two decades – in liberal democratic and authoritarian regimes alike – have resulted in more attention being paid to public opinion and stakeholder input in 'omics-related policy making. While the extent of this varies greatly by country, Levidow, Johansson, Jarvis and Richmond, Tiberghien, Howlett and Migone, Burgess, and Rehmann-Sutter all find that some movement in this direction is a common feature of current regulatory developments, and is likely to be enhanced in future years.

The third question deals with the manner in which existing regulatory trends are moving away from a 'first generation' and towards a 'next generation' pattern, and the extent to which this movement is uniform across space and time. Here the authors provide more nuanced conclusions. On the one hand, several authors see a more permissive second generation approach already in place in the US and Canada, while others find resistance to these ideas and the continuing dominance of a more precautionary approach in Europe, or the, perhaps surprising, emergence of a more precautionary approach in Asia. Lieberman and Zito argue, for example, that the stronger precautionary approach in Europe is based on the assumption of 'potential difference', rather than the 'substantial equivalence' of new biotechnologies with more conventional products, which frames approaches to regulation in the USA. Tiberghien shows that China, Japan, and Korea have all moved in a more 'European' direction and away from the permissive 'American' model as a result of civil society pressures.

Many contributors connect the complex reasons for these variations to the nature, application, and reception provided in different jurisdictions to first generation rules and regulations. In the absence of a crisis focusing public attention on agri-food technologies (such as the bovine spongiform encephalopathy (BSE) crisis in the UK and Europe in the late 1990s), the North American and Asian easing of aspects of first generation regulatory restrictions is apparent. However, where such crises have focused public attention on food safety, as occurred in some EU countries and others in Asia over the period in question, pressures for greater control and supervision have arisen (Hartley 2005; Moore 1999; Smith 2004). Kurzer and Skogstad's chapter in this volume suggests indirectly that we have good reasons to expect that these pressures will continue to be felt, as the 'potential difference' principle continues to be applied in the foreseeable future of agbiotech regulation in Europe.

By contrast, it seems likely that path dependency, a prevailing 'civic epistemology' that grants biotechnological scientists and industries privileged status in public policy debates over their regulation (Jasanoff 2005), and a continuation of current relative strengths of scientific, industrial, regulatory, and civil society stakeholders, will ensure a continuation of 'substantial equivalence' framing of North American and much Asian regulation of agricultural biotechnologies. Organised NGO and other civil society actors' campaigns concerning food safety and small farmers' viability have been decisive in pushing Chinese and Japanese authorities into a more precautionary approach with regard to labelling and product use (Tiberghien, this volume), but as Jarvis and Richmond argue in the nano-technology case, regulation in Asian countries may be driven overwhelmingly by state economic agendas. As Tiberghien shows, in the absence of suitable institutions to reconcile economic development goals with health and accountability concerns, authorities attempt local policy deliberation and other expedients to ensure regime legitimacy in the face of complex civil society opposition.

This leads directly to the fourth set of questions, concerning what drives changes to regulatory governance in this sector, and what can facilitate or block its evolution. Here the authors identify a wide range of drivers to explain variations among different countries or sectors. Significant factors highlighted in specific cases and sectors are: the 'co-production' of technology, science and society (Levidow); the different attitudes and roles of leaders and experts (Montpetit, Johansson); differences in the sources, content of, and tensions within public opinion (Weldon *et al.*); the presence or absence of prominent crises and policy failures (Kurzer and Skogstad, Howlett and Migone, Tiberghien); the dynamics of regulatory capture (Tiberghien, Levidow); states' learning and emulation (Jarvis and Richmond); NGO behaviour, sometimes in deliberative or participatory settings, with consequent impacts on the framing and politicisation of food production (Jarvis, Tiberghien, Levidow, Rehmann-Sutter); the role and attitudes of private and state business actors (Kurzer and Skogstad, Levidow, Tiberghien, Jarvis); the nature of technology or scientific advance (McHughen); international developments and factors (Zito and Lieberman); and market-based opportunity drivers versus social, reflexively need-based drivers (Rehmann-Sutter). Authors also identify key factors impeding change, including institutional arrangements such as federalism and specific government systems that promote or impede public or expert domination of policy processes (Jarvis and Richmond, Kurzer and Skogstad, Zito and Lieberman, Tiberghien). Finally, the comparative study of regulation of bio-medical technologies included in this volume raises the question of whether and how these macro-institutional differences may matter to the evolution of next generation agbiotech regulation.

The book is structured into five major sections. Preceding them, our Introduction discusses the origins of this research project and the general questions and issues motivating it. In Part 1, two chapters discuss the broad sweep of technological and regulatory developments in the field of agri-food genomics from its origins in DNA manipulation in the mid-1970s through to current and possible future developments. This discussion focuses on regulations developed in the American and European cases, where most of these techniques and technologies first emerged. Les Levidow discusses this overall record of regulation, its context of the 'co-production' of technology and society, and the issues these raise in some depth. Plant geneticist Alan McHughen then sets out the basic elements of the technologies and technological trajectories with which regulation must deal, and makes a case for regulations based on scientifically verifiable risk as opposed to politically influenced risk perceptions.

Part 2 examines the general pattern of regulatory regime development in high technology and other sectors, and the evolution of the specific pattern found in the EU. Michael Howlett and Andrea Migone examine literature that has postulated a general lifecycle of regulatory regime development, and apply this model to the agri-food genomics case. They stress the importance of examining both the 'spatial' (cross-national or cross-sectoral) and temporal variations across national or sectoral regimes, and examine the general kinds of activities and drivers associated with each stage. Anders Johansson then examines the specifics of the European case in some detail, emphasizing connections between post-1975 phases and distinctive policy models and framing assumptions. He argues that restricting the socio-economic and political considerations embedded in the evolving EU regulatory regime has not been possible, as member states and various organised interests have failed to agree on the underlying framing assumptions needed to support a more unified approach to the subject.

The three chapters in Part 3 utilise a broad comparative cross-national methodology to examine developments in several countries. Anthony Zito and Sarah Lieberman examine the EU situation discussed by Johansson, but compare it directly with the US case set out by Levidow and McHughen. Yves Tiberghien then examines three key countries in Asia – Japan, Korea, and China – and compares the nature and history of their regulatory regimes to the US and European cases. This is followed by Paulette Kurzer and Grace Skogstad's detailed examination of contemporary and potential future developments in the EU, the strongest case of continued, enhanced resistance to the promulgation of novel agri-food genomics technologies.

Part 4 analyses the agri-food case against the experiences of other similar high technology regulatory regimes. Darryl Jarvis and Noah Richmond examine the case of nano-technology in China and find that the Chinese government is determined to utilise high technology means, where it can, to advance its economy and socio-economic development. Frederic Varone, Isabelle Engeli, and Christine Rothmayr examine the closely related case of assisted-reproduction technologies (ART). They find that these, too, have moved through a multi-generational regulatory regime evolution, with different countries clustered in different stages of development.

Part 5's chapters examine the critical public opinion and participation component of the 'government to governance' hypothesis respecting the overall pattern of regulatory regime development found in many countries and sectors. Eric Montpetit provides new survey data to consider the impact of public and expert opinion on genomic technological regulation in Europe, and the clash between the two in regulatory policy development. Steven Weldon, David Laycock, Andrea Nüsser, and Colin Whelan then analyse the results of original Canadian survey data which underscore the complex divisions within the public with respect to notions of the safety and desireability of genomics applications in the agri-food sector. They propose and test a model of public opinion formation concerning biotechnological innovation that can, in principle, be applied widely both inside and outside the agbiotech sector. Michael Burgess then examines deliberative policy tools available to policy makers to constructively bring these complex divisions into the policy and regulatory processes. He focuses on the role that novel techniques for deliberative engagement can play to help reconcile some of these divisions and bring more coherence to public demands, thus better enabling governments to enhance the overall legitimacy of their rules and regulations.

In the volume's final chapter, Christoph Rehmann-Sutter re-examines philosophical and ethical assumptions of, and perspectives on, first generation agri-food genomics regulation. He argues that, to promote successful and socially beneficial applications, next generation governance of agricultural biotechnologies will have to be sensitive to market pressures while favouring need-driven development; will have to take broad public and farming community concerns about this development seriously; will attempt to develop socially reflexive relationships between genomic producers and consumers; and will need to take the social justice impacts of such applications into account while extending the scope of ethical concerns from the human world to that of the ecosystem.

Taken together, this volume updates and extends empirical and theoretical social science perspectives on earlier agricultural biotechnological regulation in ways that help to inform future policy formulation. We are confident that these chapters will stimulate debate and encourage creative thinking about continuities and paradigmatic breaks between generations of biotechnologies and regulations. On behalf of the authors, we invite readers to continue the conversation about these highly consequential issues affecting the future of these genetic technologies in sectors such as agriculture and food, and others, which are so crucial to the future of our societies.

Bibliography

Ayele, S. (2007) 'The Legitimation of GMO Governance in Africa', *Science and Public Policy* 34(4): 239–49.

Bauer, M. W. (2005) 'Distinguishing Red and Green Biotechnology: Cultivation Effects of the Elite Press', *International Journal of Public Opinion Research* 17(1): 63–89.

Bingham, N. (2008) 'Slowing Things Down: Lessons from the GM Controversy', *Geoforum* 39: 11–122.

Hartley, S. (2005) 'The risk society and policy responses to environmental risk: A comparison of risk decision-making for GM crops in Canada and the United Kingdom, 1973–2004', University of Toronto, doctoral dissertation, Department of Political Science.

Howlett, M. and A. Migone (2010a) 'Explaining Local Variation in Agri-food Biotechnology Policies: 'Green' Genomics Regulation in Comparative Perspective', *Science and Public Policy* 37(10): 781–95.

Howlett, M. and A. Migone (2010b) 'The Canadian Biotechnology Regulatory Regime: The Role of Participation', *Technology in Society* 32(4): 280–7.

Jasanoff, S. (2005a) 'Judgment Under Siege: The Three-Body Problem of Expert Legitimacy', in *Democratization of Expertise?*: 209–24. http://dx.doi.org.proxy.lib.sfu.ca/10.1007/1-4020-3754-6_12.

Jasanoff, S. (2005b) *Designs on Nature: Science and Democracy in Europe and the United States*. Princeton: Princeton University Press.

Legge, J. and R. Durant (2010) 'Public Opinion, Risk Assessment, and Biotechnology: Lessons from Attitudes toward Genetically Modified Foods in the European Union', *Review of Policy Research* 27(1): 59–76.

Levidow, L. and S. Carr (2010) *GM Food on Trial: Testing European Democracy.* New York: Routledge.

Lyall, C. (2007) 'Governing Genomics: New Governance Tools for New Technologies?' *Technology Analysis and Strategic Management* 19(3): 365–82.

Lyall, C., T. Papaioannou and J. Smith, (eds) (2009) *The Limits to Governance: The Challenge of Policy-making for the New Life Sciences.* Farnham: Ashgate.

Migone, A. and M. Howlett (2009) 'Classifying Biotechnology-Related Policy, Regulatory and Innovation Regimes; A Framework for the Comparative Analysis of Genomics Policy-Making', *Policy and Society* 28(4): 267–78.

Mintrom, M. and C. Williams (2009) 'Public Policy and Genomic Science: Managing Dynamic Change', *Policy and Society* 28(4): 253–65.

Montpetit, E. and C. Rouillard (2008) 'Culture and the Democratization of Risk Management: The Widening Biotechnology Gap Between Canada and France', *Administration & Society* 39(8): 907–30.

Moore, E. (1999) 'Science, Internationalization and Policy Networks: Regulating Genetically-Engineered Food Crops in Canada and the United States, 1973–98', University of Toronto, doctoral dissertation, Department of Political Science.

Mwale, N. (2006) 'Societal Deliberation on Genetically Modified Maize in Southern Africa: the Debateness and Publicness of the Zambian National Consultation on Genetically Modified Maize Food Aid in 2002', *Public Understanding of Science* 15(1): 89–102.

Potrykus, I. (2010) 'Lessons from the "Humanitarian Golden Rice" project: Regulation prevents development of public good genetically engineered crop products', *New Biotechnology* In Press, Accepted Manuscript. http://www.sciencedirect.com.proxy.lib.sfu.ca/science/article/B8JG4-50K5T51-1/2/469242a5ba1f0f71050d51e2bfd25b38.

Sheingate, A. (2006) 'Promotion Versus Precaution: The Evolution of Biotechnology Policy in the United States', *British Journal of Political Science* 36: 243–68.

Sjoberg L. (2005) 'Gene technology in the eyes of the public and experts: moral opinions, attitudes and risk perception', in *Working paper series in business administration*, no. 2004, vol. 7, Center for Risk Research, Stockholm School of Economics.

Skogstad, G. (2003) 'Legitimacy and/or Policy Effectiveness? Network Governance and GMO Regulation in the European Union', *Journal of European Public Policy* 10(3): 321–38.

Sleeboom, M. (2001) *Genomics in Asia. A Clash of Bioethical Interests?* London: Keegan Paul.

Smith, M. (2004) 'Mad Cows and Mad Money: Problems of Risk in the Making and Understanding of Policy', *The British Journal of Politics and International Relations* 6(3): 312–32.

Part 1

First and second generation agri-food genetic technologies and regulatory regimes: issues and overviews

2 Generating regulatory futures

Beyond agbiotech blockages to a bioeconomy?[1]

Les Levidow

Introduction: beyond biotech blockages?

Since the 1980s, agricultural biotechnology (henceforth agbiotech) has been promoted as a symbol of European progress and political-economic integration. According to proponents, agbiotech provides a clean technology for enhancing eco-efficient agro-production. By the late 1990s, however, 'GM food' became negatively associated with factory farming, its hazards, and unsustainable agriculture. GM products have generally faced commercial and/or regulatory blockages to market access in Europe. To bypass the blockage, in 2005 agbiotech was relaunched as an essential tool for the Knowledge-Based Bioeconomy (KBBE). This agenda has promoted next generation technologies for the development of non-food uses of renewable resources.

This chapter addresses several questions:

- Despite support from powerful political–economic forces in Europe, why did agbiotech encounter strong blockages there?
- What was the European conflict about?
- What can be learned from this experience for other new technologies, e.g. next generation biotech for a bioeconomy?

Co-production of technology with socio-natural order

In the 1990s European controversy over agbiotech, proponents criticised opponents for unfairly targeting or blaming a benign technology as a symbol of wider issues, such as industrial agriculture and globalisation. As this complaint illustrates, proponents often draw distinctions between a technology and its context or consequences, while critics generally emphasise links between them. Indeed, technological controversy involves power struggles over how to define the issues at stake, even the nature of the technology. These definitional issues can be illuminated by the analytical concept of co-production, which in turn will help to address the above questions.

'Co-production of technology, nature, and society' provides concepts for analysing links between the social and natural order. Science and technology can be understood as socio-technical hybrid constructs, ordering society in particular ways, while attributing that order to separate 'natural' characteristics (Jasanoff 2004: 21). In this process, people may relegate part of their experience to an apparently immutable, objective reality, as if it were given by nature.

Interactive co-production approaches study how such an order is potentially destabilised. They investigate conflicts around boundaries, for example between (un)changeable realities or natural/social realms. Interactive co-production emphasises contexts where these distinctions undergo challenge, amidst competing epistemologies. Through boundary-work, various everyday practices potentially make and stabilise those distinctions, such as between social and natural characteristics. For their authority, scientific claims depend upon such distinctions, within specific forms or changes of social order.

A co-production perspective can illuminate how technology provides a solution to a problem of socio-political order, without taking for granted a particular form (ibid: 18–19, 30). Concepts of objectivity and expertise remain important for legitimising regimes as democratically accountable. Such concepts affect how research results are taken up in public realms (e.g. as persuasive, biased, or inconclusive); how they are meant to 'solve' public problems; and how they are constructed to legitimise policy (ibid: 34). An interactive approach helps to explain how a socio-natural order is contested and changed.

At the nexus of social and natural order, there are several instruments of co-production:

- Making identities: Collective identities can help people to restore sense out of disorder, by putting things back into familiar places. These include categories and characteristics such as 'European', 'professional', 'intelligent', and so on. But these identities may also be contested or renegotiated in elaborating a different order.
- Making institutions: Institutionalised ways of knowing are reproduced (and potentially changed) in new contexts of disorder; they also serve as sites for testing or reaffirming a political culture. According to a model of market capitalism, for example, the human subject is able to form autonomous preferences, make rational choices, and act freely upon the choices so made; any exceptions are interpreted as a market failure, rather than a problem with the model.
- Making discourses: Languages are produced or modified in ways that promote tacit models of nature, society, culture, or humanity. For example, discourses may define the boundary between promising and fearsome aspects of a technology – such as links between 'un/natural' and 'un/safe' characteristics.
- Making representations: this includes the means of representation in diverse communities of practice; models of human agency and behaviour; and the uptake of scientific representations by other social actors (ibid: 38–41).

Drawing on an interactive co-production perspective, this chapter argues that agbiotech was originally promoted as a means to shape European political–economic integration around a specific socio-natural order, while also naturalising that order as an obvious future, even as an objective imperative. Civil society networks opposed and destabilised that potential new order, while also demanding alternative futures. Now agbiotech is being promoted within a wider agenda of a future bioeconomy, whose prospects likewise depend on reshaping the socio-natural order.

Making Europe safe for agbiotech

Since the 1980s, agbiotech has been promoted as a technological saviour of agro-industrial systems: GM techniques were expected to improve crops for both economic competitiveness and environmental protection. The US innovation–regulatory model was adapted by the EU as a means to make Europe safe for agbiotech, within a specific socio-natural order, as this section will show.

US neoliberal model

The global agbiotech agenda was led by the US agro-industrial complex and its government supporters. Long beforehand, these forces had turned agriculture into a rural factory of standardised commodity production, especially animal feed for global export. Transatlantic agrichemical companies developed agbiotech to further industrialise agriculture, along with the promise of alleviating its environmental damage through more efficient inputs, and replacing natural resources with knowledge embodied in novel seeds.

Promotional language attributed beneficent powers to biotechnologised nature. Key metaphors – of computer codes, commodities, and combat – were invested in nature. Molecular 'information' was decoded and edited for precisely engineered novel organisms, providing 'value-added genetics'. 'Smart seeds' would overcome threats from a wild, disorderly nature – for example, transgenes would precisely target pests and so protect crops (Levidow 1996).

In the USA, agbiotech was promoted as an imperative for new policies. These featured broader patent rights giving financial incentives to public sector research institutes as well as the private sector, alongside trade liberalisation opening foreign markets to US agri-exports and intellectual property claims. In the name of 'product-based regulation', agbiotech has been regulated under previous product legislation (e.g. for plant pests, pesticides, and so on), within existing statutory bureaucratic frameworks. Such arrangements symbolically normalised GM products as similar to conventional crops, while also minimising opportunities for public involvement.

In those ways, agbiotech was being co-produced along with neoliberal models of the natural and social order, for further commoditising knowledge. By default, in the absence of any effective means to question the innovation, the US debate focused on possible risks – whether or how GMOs could be made predictably safe for the environment. More fundamentally, nature was being redesigned and made safe for agbiotech (Sagoff 1991). Likewise, various institutional changes were making society safe for normalising and commercialising GM products.

European adaptation of US model

Since the 1980s, the European Union's integration project has likewise promoted biotech as a symbol of progress. By the early 1990s, biotech further epitomised promises of a 'knowledge-based society', promoting capital-intensive innovation as essential for economic competitiveness and thus European prosperity (CEC 1993). Through such efforts, Europe would become 'the most competitive and dynamic, knowledge-based economy in the world' (EU Council 2000).

New policies were being designed for a 'competition state', directing resources towards the domestic capacity for global competitive advantage (Cerny 1999). This included efforts to attract private sector investment, to subordinate public sector research to private sector priorities, to marketise public goods, and to generate globally competitive knowledge. Such a state was promoted through agbiotech, as both an instrument and symbol of societal progress.

The US model of agri-industrial productivity was appropriated as an inevitable European future. Soon this became linked with a neoliberal agenda emphasising economic competitiveness, projecting a unitary European interest. According to this narrative, eco-efficient technologies bring a competitive advantage and thus societal benefits, but Europe risks losing these benefits through inadequate financial rewards or over regulation. New policies sought to make Europe safe for agbiotech as normal products.

The EU agbiotech policy was also linked with a trade liberalisation agenda by invoking objective imperatives of global competition. In parallel, the European Commission promoted agbiotech as essential for economic competitiveness and thus for survival of the European agri-food sector as well as its techno–scientific capacities. By the mid-1990s, EU–US discussions were identifying 'barriers to transatlantic trade', which must be removed through regulatory harmonisation, especially for biotech products as a test case (Murphy and Levidow 2006).

EU policies also extended proprietary claims on genetic resources but provoked opposition. At issue was the concept of 'biopiracy', whether this meant unauthorised use of GM seeds, or rather 'Patents on Life' (i.e. patent rights on mere discoveries of common resources). The 'biopiracy' issue raised doubts among those on the Left and trade union groups, which were otherwise inclined to support technological innovation as societal progress. After a decade-long conflict, a 1998 EC directive extended patent rights to 'biotechnological inventions'; this broadened the scope of discoveries or techniques that could be privatised (EC 1998). As public controversy continued afterwards, several EU member states failed or refused to incorporate the Directive into national law. The European Commission brought court actions against them, but such formal trials could not harmonise national rules, nor resolve the legitimacy problem.

Research and Development (R&D) policies created greater incentives for the use of GM techniques, partly by blurring the boundary between public and private sectors. In many EU member states, public sector agricultural research institutes were allocated less state funding than before and were expected to substitute income from the private sector or from royalties on patents for GM techniques.

The European Union's R&D funding priorities complemented that shift towards marketising hitherto 'public sector' research (Levidow *et al.* 2002). As such, research institutions became more dependent on private sector financing, and critics challenged their scientific integrity regarding expert scientific advice on risk issues.

NGOs raised concerns that GM crops would stimulate agricultural intensification, undermine farmer independence, and jeopardise rural livelihoods. In response, biotech lobbies framed any socio-economic disruption, such as farmers' loss of livelihood, as a means to renew democratic societies (SAGB 1990). Thus the market was idealised as a free, naturally beneficent regulator for enhancing and allocating societal benefits.

Technicist harmonisation agenda

Through EU decision-making procedures, regulatory criteria internalised biotech-nological models of the socio-natural order. Under the EC Deliberate Release Directive, member states must ensure that GMOs do not cause 'adverse effects' (EC 1990); however, the scope of 'adverse effects' was left ambiguous, to be clarified for each product in its context. Some member states warned that GM crops could generate herbicide-tolerant weeds or pesticide-tolerant pests, but offi-cial EU risk assessments classified such effects as merely agronomic problems. This normative–regulatory judgement accepted the normal hazards of intensive monoculture, while also conceptually homogenising the agricultural environment as a production site for standard commodity crops. Through a technicist harmo-nisation agenda, Europe was being de-territorialised as a purely economic zone, devoid of cultural identities (cf. Barry 2001: 70).

Thus early EU regulatory procedures incorporated policy assumptions of the agbiotech promoters. Under 'risk-based regulation', societal decisions on agbiotech were reduced to a case-by-case approval of GM products, within a narrow definition of risks, placing the burden of evidence mainly upon the objectors. Each time the Commission proposed to authorise a GM product, it gained a qualified (two-thirds) majority in the comitology procedure representing EU member states, where dissent was marginalised.

As another controversial issue, NGOs and some member states demanded special GM labelling to ensure informed consumer choice. The European Commission initially rejected this request on several grounds: for lacking any scientific basis, unfairly impeding the internal market, and making the EU vulnerable to a US challenge under World Trade Organisation (WTO) rules. A no-labelling policy initially prevailed, despite significant dissent in all EU institutions.

All these regulations complemented the wider policy framework of higher productivity for economic competitiveness as an expected benefit from agbiotech products. This agenda was depoliticised by invoking objective imperatives such as globalisation, treaty obligations and 'risk-based regulation'. By the mid-1990s, EC policies were making Europe 'safe' for agbiotech to achieve commercial success, while subordinating regulatory criteria to economic competitiveness.

In those ways, a distinctively European approach adapted elements of the US neoliberal policy framework for agbiotech. The technology was being co-produced along with a marketisation of nature and society, in the name of eco-efficiency improvements for agriculture. Regulatory procedures authorised 'safe' GM products, which could then enter the EU internal market as extra options for farmers. They would have the free choice to buy more efficient inputs for global competitiveness. As unwitting consumers of GM food, the public would effectively support a beneficial technology serving the common good of Europe. Within this model of rational market behaviour, members of various European publics had little scope to act as citizens.

Putting agbiotech on trial, reshaping regulations

By promoting agbiotech within a neoliberal framework, the EU system provoked great suspicion and even opposition, which grew from the mid-1990s onwards. Agbiotech was turned into a symbol of anxiety about multiple threats: the food chain, agro-industrial methods, their hazards, state irresponsibility, and political unaccountability through globalisation. The controversy often gained large public audiences through the mass media, as well as active involvement of many civil society groups. They took up concepts from small activist groups as well as from high profile campaigns of large NGOs. Together these activities developed citizens' capacities to challenge official claims and created civil society networks to which governments could be held accountable.

These activities criticised, used, and eventually reshaped the EU regulations. Demands for accountability took the form of various formal and informal trials. These dynamics continuously expanded trials, defendants, and arenas – what was put on trial, how, where, and by whom. Such trials arose along three overlapping themes – safety versus precaution, eco-efficiency versus agro-industrial hazards, and globalisation versus democratic sovereignty – as shown in this section. (For more details and sources, see Levidow and Carr 2010.)

Safety claims versus precaution

Lab and field trials were intended to generate evidence of product safety, thus demonstrating a scientific basis for expert risk assessments, which in turn could justify commercial authorisation of GM products. Yet safety science became contentious. Expert safety claims underwent criticism for bias, ignorance, and optimistic assumptions. Such criticism gained force from suspicion that public sector scientists had lost any independence from agbiotech promotion.

When France led the EU-wide approval of Bt maize 176, its favourable risk assessment was widely criticised by member states as well as NGOs. When France further proposed to approve maize varieties derived from Bt 176 in 1998, Ecoropa and Greenpeace filed a challenge at the *Conseil d'Etat* (the French administrative high court) on several grounds: that the risks had not been properly assessed, that the correct administrative procedures had not been followed, and that the Precautionary Principle had not been properly applied. These NGO arguments gained some support in the court's interim ruling. Thus, a government was judicially put on trial for failing to put a GM product on trial in a rigorous way.

When UK lab experiments claimed to find harm to rats from GM potatoes, the disclosure led to trials of other kinds. The project leader, Arpad Pusztai, questioned the safety of GM foods on a television programme. He was soon dismissed from his post and was then subjected to character assassination by other scientists. His experimental methods were criticised by a Royal Society report. International networks of scientists took opposite sides on that issue. NGOs put his employers and other persecutors symbolically on trial, by attributing their actions to political and commercial motives (Levidow 2002; Levidow and Carr 2010: 100–02).

When a Swiss lab experiment found that an insecticidal Bt maize harmed a beneficial insect (lacewing), expert authority was put on trial. Criticising the experiment, other scientists cast doubt on its methodological rigour and its relevance to commercial farming as grounds to discount the results in the regulatory arena. In response, agbiotech critics reversed the accusation: they raised similar doubts about the rigour of routine experiments that had supposedly demonstrated safety. Potential harm to non-target insects remained a high profile issue, attracting further research and expert disagreements. Citing scientific uncertainties, some regulatory authorities rejected Bt maize or demanded that its cultivation be subject to special monitoring requirements at the commercial stage, thus further testing safety claims and elaborating test protocols (Levidow and Carr 2010: 182–3).

In the latter two risk issues, surprising experimental results were deployed to challenge safety claims, optimistic assumptions, and expert safety advice. When new evidence of risk was criticised for inadequate rigour or relevance to realistic commercial contexts, similar criticisms were raised against safety claims and their methodological basis. Regulatory authorities were put symbolically on trial for failure to develop adequate scientific knowledge for risk assessment, instead depending on companies for test data.

For the safety assessment of GM food, EU regulatory procedures and criteria likewise were put on trial. Under the EU's Novel Food Regulation, for example, GM products could be approved via a simplified procedure in cases where they had substantial equivalence with a non-GM counterpart. After such approval decisions about several foods derived from GM maize, Italy banned them partly on grounds that the decisions had inadequate scientific evidence to demonstrate substantial equivalence. The Commission sought to lift the ban and so requested support from the EU regulatory committee of member states in 2000, thus putting Italy on trial by its peers. But they instead sided with Italy, while also criticising the regulatory shortcut under the Novel Food Regulation (Standing Committee, 2000).

After this role reversal, the Commission abandoned substantial equivalence as a statutory basis for easier approval of novel foods (EC 2003a). In risk assessment procedures, substantial equivalence continued as a 'comparative assessment'; this was broadened to encompass more methodological issues, scientific uncertainties, and types of scientific evidence (Levidow *et al.* 2007). Such comparison with conventional products has remained contentious among member states as well as civil society groups.

Globalisation versus democratic sovereignty

Given that agbiotech promoters emphasised globalisation as an imperative for GM products, critics could portray them as a threat and agent of 'globalisation'. Since the mid-1990s, field trials have been meant to demonstrate the agronomic efficacy and safety of GM crops, as well as the diligent responsibility of the authorities in avoiding any environmental harm. However, the fields were turned into theatrical stages for protest. They used an 'X' or biohazard symbol to cast agbiotech as pollutants and unknown dangers, thus justifying sabotage as environmental protection. When facing prosecution, activists used the opportunity to put the state symbolically on trial for inadequately evaluating or controlling GM crops, as a failure of responsibility.

Activists appealed to democratic sovereignty when carrying out and defending sabotage actions on field trials. The UK government implied that decisions about GM crops lay elsewhere, beyond its political control. This claim was denounced as an irresponsible, undemocratic surrender to globalisation. As a response to deferential regulatory decisions, such as the UK government's above, opponents defended sabotage as democratic accountability. Further to the French example above, in 1998 the WTO approved higher US tariffs against several specialty foods including Roquefort cheese, as compensation for lost exports of US beef. *Paysan* activists attacked McDonalds as a symbol of WTO rules forcing the world to accept hazardous *malbouffe* such as hormone treated beef and GM food. As defendants in court, *paysans* sought to put 'globalisation' on trial, represented by the French government as well as biotech companies.

Democratic sovereignty also became an explicit theme in judicial trials and regulatory procedures. When some EU member states explicitly refused to support authorisation of any more GM products from 1999 onwards, they were demanding precautionary reforms in EU rules and regulatory criteria. At the same time, this defiance was turned into a public symbol of European sovereignty versus globalisation driven by the USA.

Democratic sovereignty became general grounds to justify measures or actions restricting GM products at the national or regional level. By the late 1990s, fewer member states were willing to support Commission proposals to approve new GM products. Some signed formal statements that they would refuse to do so. Lacking a qualified majority, in 1999 the EU Council effectively suspended the decision-making procedure for new GM products; this move was widely called the *de facto* moratorium. Meanwhile some member states also banned GM products that had gained EU-wide approval (Levidow *et al.* 2000).

The European Commission faced a dilemma: either ignore official procedures, or else authorise the products anyway without political legitimacy. To go beyond this dilemma, the Commission proposed more stringent criteria for risk assessment and market stage monitoring in a revised Directive, which was eventually adopted (EC 2001). Although these more precautionary criteria were meant to accommodate dissent and so facilitate regulatory decisions, assessment criteria remained contentious after the EU-wide procedure resumed in 2003.

'Globalisation' also framed conflicts over GM labelling. The originator of GM soya, Monsanto, was denounced by various NGOs as a global bully 'force-feeding us GM food'. Before the European Commission approved GM soya in 1996, NGOs and some member states demanded mandatory labelling for all GM foods. However, this demand was rejected, with warnings that any such requirement would provoke a WTO case against the EU.

On this basis, the no-labelling policy became vulnerable to attack as globalisation undermining consumer choice and democratic sovereignty. Local protests at supermarkets demanded GM labelling and non-GM alternatives, in campaigns linked with Europe wide consumer and environmentalist groups. By 1998, European retail chains adopted voluntary labelling of their own brand products with GM ingredients. Companies variously labelled their products as 'contains GM' or as 'GM-free', in compliance with different criteria established by EU member states. Meanwhile NGOs carried out surveillance of GM material in food products, some not labelled 'GM', in order to protest against them and to warn consumers.

Together these regulatory inconsistencies and protests potentially destabilised the EU's internal market for processed food products. So the EU established more comprehensive standard criteria; these went beyond detectability and so required an audit trail of paper documentation. Eventually EU law required comprehensive GM labelling and traceability of GM material (EC 2003b), encompassing a broader range of products than before.

GM products also faced a commercial boycott. By the late 1990s, all European supermarket chains excluded GM ingredients from their own brand products, rather than label them as 'GM'; some mentioned precaution and/or consumer choice as reasons. By now GM ingredients were relegated to animal feed from two main sources: imported GM soya was still used in some animal feed, though some suppliers advertised 'GM-free' meat or poultry; Bt insecticidal maize was (and still is) widely cultivated in Spain, where nearly all maize enters a common supply chain for animal feed. This agro-industrial system resembles the USA's. These exceptions prove the rule of agbiotech being co-produced with models of the socio-natural order.

Eco-efficiency versus agro-industrial hazards

Agbiotech began with a cornucopian promise. With precisely controlled genetic changes, GM crops would provide smart seeds as eco-efficient tools for sustainably intensifying industrial agriculture. These promises were extended by the 'Life Sciences' project, featuring mergers between agro-supply and pharmaceutical companies, in search of synergies between their R&D efforts. Its narrative promised health and environmental benefits as solutions to general societal problems.

Critics turned agbiotech into a symbol of multiple threats. Productive efficiency was pejoratively linked with agro-industrial hazards; for example, the epithet 'mad soja' drew analogies to the BSE epidemic. Biotech companies were accused of turning consumers into human guinea pigs.

Through politically constituted cultural meanings, agbiotech was put symbolically on trial as an unsustainable, dangerous, misguided path. In France, critics cast agbiotech as *malbouffe* (junk food), as threats to high quality *produits du terroir*. In Italy, GM crops were cast as agro-industrial competition and 'uncontrolled genetic contamination', threatening diverse, local quality agriculture. Using the term *Agrarfabriken* (factory farm), German critics linked agbiotech with intensive industrial methods, threatening human health, the environment, and agro-ecological alternatives. Institutions faced greater pressure to test claims that GM crops would provide agro-environmental improvements as well as safety.

Those informal trials shaped conflicts over regulatory criteria from the mid-1990s onwards. When EU procedures initially evaluated GM crops for cultivation purposes, they were deemed safe by accepting the normal hazards of intensive monoculture. This normative stance was portrayed as a scientific judgement, while casting any criticism as irrelevant or political. Yet such hazards were being highlighted by critics, framing risks in successively broader ways. Their discourses emphasised three ominous metaphors: 'superweeds' leading to a genetic treadmill, thus aggravating the familiar pesticide treadmill; broad-spectrum herbicides inflicting 'sterility' upon farmland biodiversity; and pollen flow 'contaminating' non-GM crops.

These ominous metaphors expanded the charge sheet of hazards for which GM products were kept on trial. Moreover, these broader hazards would depend on the behaviour of agro-industrial operators, which consequently became a focus of prediction, discipline, and testing. Regulatory procedures came under pressure to translate the extra hazards into risk assessments. In its risk assessment for GM herbicide-tolerant oilseed rape, Bayer claimed that farmers would eliminate any resulting herbicide-tolerant weeds and so avoid weed control problems, but Belgian experts questioned the feasibility of such measures. Citing that advice, the Belgian national authority rejected the proposal to authorise cultivation uses, rather than invite the company to test extra hazards. So a proposal went forward only for food and feed uses, gaining EU approval on that limited basis (EC 2007).

GM herbicide-tolerant crops had been promoted as a means to reduce herbicide usage and thus to protect the environment. But UK critics portrayed more efficient weed control as a hazard: broad-spectrum herbicides could readily extend the 'sterility' of greenhouses to the wider countryside, which would be turned into 'green concrete'. The UK government was widely criticised for ignoring the agro-environmental implications. The Environment Ministry eventually took responsibility and funded large-scale field experiments, to simulate and thus predict farmer behaviour in spraying herbicides. These trials were meant to facilitate the 'managed development' of such crops. But experimental results indicated potentially greater harm from some GM crops than their conventional counterpart (Champion *et al.* 2003). These results led to a regulatory impasse for GM crops that could have been approved by the UK. Through a more precautionary regulatory procedure, agro-industrial efficiency was cast as an environmental threat to be investigated and avoided.

From the UK controversy in particular, the EU system underwent pressure to broaden the potential effects and their causes that warrant evaluation. The *de facto* EU moratorium led to a revised EC directive, which broadened risk assessment criteria to encompass any changes in agricultural management practices, such as in herbicide spraying, as well as indirect and long term effects (EC 2001). This broader scope potentially accommodated dissent into regulatory procedures, but public and expert debate continuously questioned safety assumptions. Broader accounts of harm meant greater uncertainty about whether GM crops could generate such harm in the agro-food chain, so risk assessments needed to anticipate human practices as well as their environmental effects.

Co-production of biotechnologised nature

In sum, an interactive co-production perspective can illuminate the early strategies of promoting agbiotech for a new socio-natural order, which was eventually destabilised by opponents. Agbiotech had been originally promoted as a clean technology enhancing natural properties. Through precise genetic changes, GM crops would efficiently use natural resources to combat plant pests and to minimise agrochemical usage, thereby developing sustainable agriculture. Such beneficent claims were challenged along several lines: safety versus precaution, eco-efficiency versus agro-industrial hazards, and globalisation versus democratic sovereignty. The entire development model – now called 'GM food', or *OGM* in Romance languages, or *Gen-Müll* (garbage) in German, etc. – was negatively associated with factory farming, its health hazards, and unsustainable agriculture. The would-be new order was stigmatised as an abnormal, dangerous disorder.

Agbiotech was turned into a symbol, object, and catalyst for multiple overlapping trials. The defendant that was symbolically on trial expanded from product safety, to biotech companies, their innovation trajectory, regulatory decision making, expert advisors, and government policy. Europe was told that it had no choice but to accept agbiotech, yet this imperative was turned into a test of democratic accountability for societal choices. In these ways, protest challenged the democratic legitimacy of a biotech-driven development pathway, as well as a European integration model for further commoditising natural resources and redesigning agriculture accordingly.

Opposition activities criticised, used, and eventually reshaped EU regulations. These were originally meant to marginalise citizens' involvement or to accommodate public concerns, in ways facilitating an internal market for agbiotech products, but instead the regulatory framework itself became more contentious. By the late 1990s, agbiotech was being co-produced with representations of biotechnologised nature as suspect, potentially abnormal, and warranting continuous surveillance.

Designing next generation biotech for a bioeconomy

In the late 1990s, when European protest led to blockages of GM products, European plant science faced reductions in research funds from both state and private sector sources. As a way forward, agbiotech was relaunched for a Knowledge-Based Bioeconomy (DG Research 2005). This KBBE vision extends mechanical and informatic metaphors from earlier biotech; cells become factories or microcomputers, especially as a basis for linking agriculture with the chemical industry.

> ... biotech employs microorganisms, such as yeasts, moulds and bacteria as so-called 'cell factories' and enzymes to produce goods and services. This implies developing and producing chemicals at the cellular level by exploiting and adjusting natural processes in living organisms to generate the substances and enzymes needed by industry (DG Research 2005: 9).

Soon the KBBE was officially defined as 'the sustainable, eco-efficient transformation of renewable biological resources into health, food, energy and other industrial products' (DG Research 2006: 3). In the Food, Agriculture, Fisheries, and Biotechnology (FAFB) programme of the EU's Framework Programme 7, research agendas have been shifted towards non-food uses of renewable resources. As a novel development, entire systems are being redesigned for horizontally integrating agriculture with other industrial sectors, especially through molecular level decomposition and recomposition of natural resources (Levidow 2011). Let us examine how this agenda potentially co-produces technology, nature and society.

Efficient techno-fixes for resource constraints

EU level research agendas have been driven by European Technology Platforms (ETPs). Invited by the European Commission, ETPs were meant to define research agendas that would attract industry investment, especially as means to fulfil the Lisbon Agenda goal of three per cent GDP (gross domestic product) being spent on research (EU Council 2000; DG Research 2004). ETPs were mandated to involve 'all relevant stakeholders' in developing a 'common vision' emphasising societal needs and benefits. For the agro-food-forestry-biotech sectors, now seen as the KBBE, ETPs were initiated mainly by industry lobby organisations, with support from scientist organisations and COPA (Comité des Organisations Professionnelles Agricoles), representing the relatively more industrialised farmers.

The KBBE concept nearly equates sustainable development with more efficiently using renewable resources. In the dominant account, such resources become biomass or raw materials as interchangeable inputs into an industrial process. For example, the KBBE is 'the sustainable production and conversion of biomass into various food, health, fibre and industrial products and energy'; such conversion is also sustainable, being efficient, producing little or no waste, and often using biological processing, according to a consortium of ETPs (Becoteps 2011: 5). Likewise, agriculture must provide 'competitive raw materials', according to a report for the EU Presidency (Clever Consult 2010). In this new vision, agriculture provides raw materials that can be broken down into various components for further processing.

Eco-efficiency is sought in novel inputs, outputs and processing methods. Research seeks generic knowledge for identifying substances that can be extracted, decomposed, and recomposed. From this baseline, more specific knowledge can be privatised: 'knowledge and intellectual property will be critical...' (Plants for the Future TP 2007: 9). Farmers become (or remain) purchasers of 'efficient' inputs, such as novel crops for enhancing soil fertility and thus productivity.

These remedies correspond to specific accounts of societal needs, whereby agriculture supplies biomass as raw materials for commodity markets. Along with food security, 'biomass as a renewable raw material for industry will be the basis of the coming integrated Bioeconomy' (Becoteps 2011: 5). In the dominant narrative, greater pressure on natural resources, and thus food insecurity, comes from global market demand. For example, 'the worldwide demand for feed will increase dramatically as a result of the growing demand for high value animal protein' (Plants for the Future TP 2007: 3). Somehow, the increasing demand remains exogenous to the agro-industrial production system, which must accommodate the demand sustainably, i.e. more efficiently through technological innovation. For example, 'In the coming decades, we anticipate the creation of more efficient plants (able to use water and fertiliser more efficiently and to be self-resistant to pests), leading to more efficient farms and new economic opportunities' (Plants for the Future TP 2007: 5, 9).

As a specific model of the socio-natural order, then, greater efficiency is attributed to capital-intensive inputs (e.g. cell factories), as a basis for an expanding commodities market to become more sustainable and to enhance food security. Resource constraints can be turned into new commercial opportunities, while the agro-supply system can avoid responsibility for the greater demands on resources. Here sustainability means eco-efficient productivity through resources that are renewable, reproducible and, therefore sustainable (Birch *et al.* 2010).

Mining agriculture as new oil wells

To optimise renewable biological resources, molecular level techniques become essential tools for identifying and validating compositional characteristics. As crucial knowledge, systems biology will predict effects of new genetic combinations:

> Systems biology will reveal how natural genetic variation creates biodi-
> versity and, together with innovative genomic technologies, will cause
> a paradigm shift in how we breed plants in the future. It will replace trial
> and error with targeted and predictive breeding to deliver desired new

traits and varieties… As systems biology requires massive quantitative genome-wide data, technologies – such as protein arrays – need to be developed to analyse simultaneously numerous possible parameters at multiple time points (Plants for the Future, 2007: 63, 66).

In this KBBE vision, next generation agbiotech must be integrated with converging technologies, that is, integrated with infotech and nanotech. These priorities were incorporated into the FAFB programme:

> Research will include 'omics' technologies, such as genomics, proteomics, metabolomics, and converging technologies, and their integration within systems biology approaches, as well as the development of basic tools and technologies, including bioinformatics and relevant databases, and methodologies for identifying varieties within species groups (DG Research 2006: 12).

Together these techniques link compositional characteristics with market opportunities, which are anticipated as value chains, a concept which helps to mobilise new commercial partnerships. In the dominant vision, technological–industrial innovation must horizontally integrate the agriculture and energy sectors; 'the production of green energy will also face the exceptional challenge of global industrial restructuring in which the very different value chains of agricultural production and the biorefining industries must be merged with the value chains of the energy providers' (Plants for the Future TP 2007: 33).

The search for lignocellulosic fuels illustrates how global market opportunities frame technical problems. As an evolutionary feature, lignin in plant cell walls impedes their breakdown and protects them from pests. From the standpoint of cross-sectoral molecular level integration, however, lignin limits the use of the whole plant as biomass for various uses including energy. For agricultural, paper, and biofuel feedstock systems 'lignin is considered to be an undesirable polymer' (EPOBIO 2006: 27).

To overcome the limitation, plants are redesigned for new value chains linking agriculture with energy. 'This larger-scale research effort was considered essential to achieve the foundation for designing *in planta* strategies to engineer bespoke [custom-made] cell walls optimised for integrated biorefinery systems' (EPOBIO 2006: 34). GM techniques are used to modify the lignin content of wood, for example, 'to improve pulping characteristics by interfering with lignin synthesis' (ibid; Coombs 2007: 55).

More generally, plants are redesigned for a biomass processing industry. An 'integrated diversified biorefinery' would use renewable resources more efficiently via more diverse inputs and outputs, which can be flexibly adjusted according to global market prices. As an ideal of eco-efficiency, closed-loop recycling successively turns wastes into raw materials for the next stage. Agriculture becomes a biomass factory; residues become waste biomass for industrial processes. Horizontal integration is being promoted through new commercial linkages between novel crops, enzymes, and processing methods.

In this vision, agriculture becomes future 'oil wells'. An international conference on the biorefinery brought together diverse industries with a common aim to integrate biomass sources and products:

> Participants included members of the forestry, automotive, pulp and paper, petroleum, chemicals, agriculture, financial, and research communities.... It was noted by DOE and EU that both the U.S. and EU have a common goal: Agriculture in the 21st century will become the oil wells of the future – providing fuels, chemicals and products for a global community (BioMat Net 2006).

'Oil well' provides an appropriate metaphor for this agenda of genetically modifying plants and/or enzymes for conversion into various industrial products. Organisms become interchangeable raw materials to be 'cracked' like oil. This concept has been elaborated through compositional analogies to crude oil, thus reordering nature; 'New developments are ongoing for transforming the biomass into a liquid 'biocrude', which can be further refined, used for energy production or sent to a gasifier' (Biofrac 2006: 21).

In the name of a common societal vision then, the dominant KBBE agenda potentially co-produces technology, nature, and society along lines further commoditising resources. In this vision, converging technologies will unlock the beneficent natural properties of plants. Eco-efficient inputs will link environmental and economic sustainability.

Alongside that dominant account, others contend for influence. In an agro-ecological account of the KBBE, ecological processes enhance and integrate eco-efficiency. 'Organic farming is a highly knowledge-based form of agriculture involving both high tech and indigenous knowledges and is based on the farmer's aptitude for autonomous decision making' (Niggli *et al.* 2008: 34). Organic farming attempts to keep cycles as short and as closed as possible, in order to use biodiverse resources more efficiently. These practices enhance resource efficiency by enhancing internal inputs as substitutes for external inputs, while also maximising outputs. Residues are seen as media for recycling nutrients via ecological processes and so replenishing soil fertility (Schmid *et al.* 2009). Although this account remains marginal, it highlights the societal choices implicit in the dominant one (Levidow *et al.* 2012).

Conclusions: lessons for future technology?

Let us return to the initial questions: Why did agbiotech encounter strong blockages in Europe? What can be learned from this experience for other new technologies?

Agbiotech was initially co-produced with nature and society within a neoliberal globalisation framework. In turn, agbiotech was used to promote those policies by remaking discourses, institutions, and identities. As a resource for GM crops, nature was invested with metaphors of codes, combat, and commodities. In the mid-1990s, EU regulatory criteria internalised assumptions of agbiotech innovation by, for example, accepting the normal hazards of intensive monoculture, while also welcoming greater efficiency as an environmental benefit. Through a technicist harmonisation agenda, the European environment was conceptually homogenised for political–economic integration. Europe was being made safe for agbiotech.

GM products were reaching the commercial stage in a period when food hazards were widely attributed to agri-industrial methods, their profit driven efficiency, deregulatory policies, and official expert ignorance. So criticism of agbiotech resonated strongly with public anxieties. Agbiotech also acquired public meanings through EU policy frameworks: regulatory harmonisation, trade liberalisation, and commoditisation of plant genetic resources. Together those policy frameworks provided a vulnerable target, thereby linking various opponents of agbiotech, agro-industrial development, and neoliberal globalisation. This conflict was illuminated by an interactive co-production approach to strategies for imposing a specific future as an objective imperative, in turn provoking a broad opposition and demands for alternatives.

More recently, to bypass blockages of GM agri-food products, agbiotech has been promoted for non-food uses in a future European bioeconomy. This again invokes economic competition and natural characteristics (e.g. biocrude) as an objective basis for European political–economic integration along specific lines. In the name of overcoming constraints on natural resources, this would horizontally integrate agriculture with industry for global value chains in proprietary knowledge. European prospects for next generation agbiotech depend on reshaping the socio-natural order according to such a bioeconomy.

From the experience of the European agbiotech controversy in the 1990s, commentators have drawn wider lessons, including some dubious ones. For example, 'The easiest way for the nanotechnology community to avoid the problems experienced in the deployment of biotechnology is to provide accurate information and encourage critical, informed analyses.' (McHugen 2008: 51) This attributes the earlier public controversy to a deficit of publicly available information, yet its reliability and accuracy were contested in a context where greater knowledge generally led to greater opposition.

Another lesson often heard was that the next novel technology could become 'another GM' if the public is not adequately consulted at an early stage. Conversely, it is also said that greater public involvement or deliberation could help to avoid societal conflict over technological innovations. For example, 'Given the opportunity to deliberate on such innovations, the public voice can be expected to be measured and moderate' (Gaskell 2008: 257).

In their own way, those two distinct lessons each decontextualise technology from its political–economic agendas. From the 1990s agbiotech conflict, we could draw different lessons, albeit less comfortable ones:

- technology is always co-produced with a specific form of the socio-natural order, thus pre-empting other choices of societal future;
- societal conflict arises from such non-choices; and so
- technology, information, and even deliberation cannot remain credibly neutral in relation to those choices.

In sum, an interactive co-production perspective helps to illuminate the 1990s conflict over agbiotech and the possible lessons for future technoscientific developments. Europe was told that it must accept agbiotech, whose design and policy context potentially naturalised a specific future society, as it objectively required. Yet this supposed imperative was turned into a test of democratic accountability for societal choices.

Therefore, prospects for avoiding 'another GM' controversy – or perhaps for creating one – depend upon how a technological innovation models the socio-natural order and how state bodies attempt to promote that order. If a political–economic choice is represented as an objective imperative, then such an innovation may be successfully naturalised, stabilised, and imposed. Alternatively, civil society opposition may destabilise and block that political–economic choice, so that the innovation can be co-produced (if at all) along different lines. Those potential outcomes pose either a threat or an opportunity, depending on one's aims.

Note

1 This chapter draws upon two research projects: 'Precautionary Expertise for GM Crops' (PEG) funded by the European Community's Framework Programme 5, Quality of Life programme, socio-economic aspects, grant agreement no. QLG7-2001-00034, during 2002–04; and 'Co-operative Research on Environmental Problems in Europe' (CREPE), funded by the European Community's Framework Programme 7, Science in Society programme under grant agreement no. 217647, during 2008–10.

Bibliography

Barry, A. (2001) *Political Machines: Governing a Technological Society*. London: Athlone.

Becoteps (2011) *The European Bioeconomy in 2030: Delivering Sustainable Growth by addressing the Grand Societal Challenges*, www.becoteps.org, http://www.plantetp.org/images/stories/stories/documents_pdf/brochures_web.pdf, accessed 28 February 2011.

Biofrac (2006) *Biofuels in the European Union: A Vision for 2030 and Beyond*. Final draft report of the Biofuels Research Advisory Council.

BioMat Net (2006) 1st International Biorefinery Workshop.

Birch, K., Levidow, L. and Papaioannou, T. (2010) 'Sustainable capital? The neoliberalization of nature and knowledge in the European "Knowledge-Based Bioeconomy"', *Sustainability* 2(9): 2898–918, http://www.mdpi.com/2071-1050/2/9/2898/pdf, accessed 28 February 2011.

CEC (1993) *Growth, Competitiveness and Employment: The Challenges and Ways Forward into the 21st Century*. Brussels: Commission of the European Communities.

Cerny, P. G. (1999) 'Reconstructing the political in a globalizing world: states, institutions, actors and governance', in F. Buelens, (ed.) *Globalization and the Nation-State*. Cheltenham: Edward Elgar (89–137).

Champion, G. T., May, M. J., Bennett, S., Brooks, D. R., Clark, S. J., Daniels, R. E., Firbank, L. G., Haughton, A. J., Hawes, C., Heard, M. S., Perry, J. N., Randle, Z., Rossall, M. J., Rothery, P., Skellern, M. P., Scott, R. J., Squire, G. R. and Thomas, M. R. (2003) 'Crop management and agronomic context of the Farm Scale Evaluations of genetically modified herbicide-tolerant crops', *Philosophical Transactions: Biological Sciences*, Series B 358(1439): 1801–18.

Clever Consult (2010) Report for the EU Presidency, *The Knowledge-Based Bio-Economy (KBBE) in Europe: Achievements And Challenges*. Clever Consult BVBA, www.cleverconsult.eu, accessed 28 February 2011.

Coombs, J. (2007) *Building the European Knowledge-Based Economy: The Impact of 'Non-food' Research (1988 to 2008)*. CPL Scientific Publishing.

DG Research (2004) *Technology Platforms: From Definition to Implementation of a Common Research Agenda*. Brussels: Commission of the European Communities.

DG Research (2005) *New Perspectives on the Knowledge-Based Bio-Economy: conference report*. Brussels: Commission of the European Communities.

DG Research (2006) FP7 Theme 2: Food, Agriculture, Fisheries and Biotechnology (FAFB), 2007 work programme. Brussels: Commission of the European Communities.

EC (1990) 'On the deliberate release into the environment of genetically modified organisms 35' *Council Directive 90/220/EEC*, 23 April 1990, OJ L 117, 8 March 1990: 15 ff.

EC (1997) Regulation 258/97/EC of 27 January 1997 concerning novel foods and novel food ingredients, *Official Journal of the European Union*, L 43, 14 February: 1–6.

EC (1998) Directive 98/44/EC of the European Parliament and of the Council on Protection of Biotechnological Inventions, *Official Journal of the European Union*, 30 July, L 213: 13.

EC (2001) European Parliament and Council Directive 2001/18/EC of 12 March on the deliberate release into the environment of genetically modified organisms and repealing Council Directive 90/220/EEC, *Official Journal of the European Union* L 106: 1–38.

EC (2003a) Regulation 1829/2003 of 22 September 2003 on genetically modified food and feed, *Official Journal of the European Union*, 18 October, L 268: 1–23.

EC (2003b) Regulation 1830/2003 of 22 September 2003 concerning the traceability and labelling of GMOs and traceability of food and feed produced from GMOs and amending Directive 2001/18, *Official Journal of the European Union*, 18 October, L 268: 24–28.

EC (2007) Commission Decision 2007/232/EC of 26 March 2007 concerning the placing on the market, in accordance with Directive 2001/18/EC of the European Parliament and of the Council, of oilseed rape products (*Brassica napus L.*, lines Ms8, Rf3 and Ms8xRf3) genetically modified for tolerance to the herbicide glufosinate-ammonium (notified under document number C(2007) 1234), *Official Journal of the European Union*, 17 April, L 100: 20.

EPOBIO (2006) *Products From Plants – The Biorefinery Future. Outputs from the EPOBIO Workshop*, Wageningen, 22–24 May 2006, http://www.epobio.net/workshop0605.htm, accessed 28 February 2011.

EU Council (2000) *An Agenda of Economic and Social Renewal for Europe: (aka Lisbon Agenda)*. Brussels, European Council, DOC/00/7.

Gaskell, G. (2008) 'Lessons from the bio-decade', in K. David and P. B. Thompson (eds), *What Can Nanotechnology Learn From Biotechnology?: Social and Ethical Lessons for Nanoscience from the Debate Over Agrifood Biotechnology and GMOs*. London: Academic Press (237–58).

Jasanoff, S. (2004) 'Ordering knowledge, ordering society', in S. Jasanoff, (ed.) *States of Knowledge: The Co-production of Science and Social Order*. London: Routledge (13–45).

Levidow, L. (1996) 'Simulating Mother Nature, industrializing agriculture', in G. Robertson *et al.* (eds), *FutureNatural: Nature, Science, Culture*. London: Routledge (55–71).

Levidow, L. (2002) 'Ignorance-based risk assessment? Scientific controversy over GM food safety', *Science as Culture* 11(1): 61–7.

Levidow, L. (2011) (ed.) *Agricultural Innovation: Sustaining What Agriculture? For What European Bio-Economy?*, from an FP7 project, Co-operative Research on Environmental Problems in Europe (CREPE), www.crepeweb.net, accessed 28 February 2011.

Levidow, L. and Carr, S. (2010) *GM Food on Trial: Testing European Democracy*. London: Routledge.

Levidow, L., Carr, S. and Wield, D. (2000) 'Genetically modified crops in the European Union: regulatory conflicts as precautionary opportunities', *Journal of Risk Research* 3(3): 189–208.

Levidow, L., Søgaard, V. and Carr, S. (2002) 'Agricultural PSREs in Western Europe: research priorities in conflict', *Science and Public Policy* 29 (4): 287–95.

Levidow, L., Carr, S. and Wield, D. (2005) 'EU regulation of agri-biotechnology: precautionary links between science, expertise and policy', *Science & Public Policy* 32(4): 261–76, http://technology.open.ac.uk/cts/peg/sppaug2005eu%20fin.pdf

Levidow, L., Murphy, J. and Carr, S. (2007) 'Recasting "Substantial Equivalence": transatlantic governance of GM food', *Science, Technology and Human Values* 32(1): 26–64.

Levidow, L., Birch, K. and Papaioannou, T. (2012) Divergent paradigms of European agro-food innovation: The Knowledge-Based Bio-Economy (KBBE) as an R&D agenda, *Science, Technology and Human Values* 37, forthcoming.

McHugen, A. (2008) 'Learning from mistakes', in K. David, and P. B. Thompson (eds), *What Can Nanotechnology Learn From Biotechnology? Social and Ethical Lessons for Nanoscience from the Debate Over Agrifood Biotechnology and GMOs*. London: Academic Press (33–53).

Murphy, J. and Levidow, L. (2006) *Governing the Transatlantic Conflict over Agricultural Biotechnology: Contending Coalitions, Trade Liberalisation and Standard Setting*. London: Routledge.

Niggli, U., Slabe, A., Schmid, O., Halberg, N. and Schlüter, M. (2008) *Vision for an Organic Food and Farming Research Agenda to 2025*, Brussels: Technology Platform Organics, http://www.organic-research.org/index.html, http://orgprints.org/13439, accessed 28 February 2011.

Plants for the Future TP (2007). *European Technology Platform Plants for the Future: Strategic Research Agenda 2025. Part II*. Brussels: European Plant Science Organisation.

SAGB (1990) Senior Advisory Group on Biotechnology, *Community Policy for Biotechnology: Priorities and Actions*. Brussels: European Chemical Industry Council (CEFIC).

Sagoff, M. (1991) 'On making nature safe for biotechnology', in L. Ginzburg (ed.) *Assessing Ecological Risks of Biotechnology*. Stoneham, MA: Butterworth-Heineman (341–65).

Schmid, O., Padel, S., Halberg, N., Huber, M., Darnhofer, I., Micheloni, C., Koopmans, C., Bügel, S., Stopes, C., Willer, H., Schlüter, M. and Cuoco, E. (2009) *Strategic Research Agenda for Organic Food and Farming*, Brussels: Technology Platform Organics, http://www.organic-research.org/index.html.

Standing Committee (2000) Summary record of 78th meeting, EU Standing Committee on Foodstuffs, 18–19 October, http://ec.europa.eu/comm/food/fs/rc/scfs/rap02_en.html, accessed 28 February 2011.

3   Learning from experience

How do we use what we have learned to reform regulatory oversight of new agricultural biotechnologies?

Alan McHughen

Genetic engineering (GE) or genetic modification (GM) technologies based on recombinant DNA (rDNA) arose in the 1970s and provided a wide range of products, including pharmaceuticals, industrial compounds, crops, and foods. The biosafety concerns related to these technologies and their products were expressed early on by the very scientists pioneering the technology (Berg *et al*. 1974). As a result, biosafety regulations were quickly promulgated in various countries to address the legitimate safety concerns regarding rDNA and its products to both human health and the environment.

Western countries now have almost forty years of experience dealing with genetically modified organisms (GMOs) as the products of rDNA technology, including fifteen years of GM crops in commercial production. However, most of the biosafety regulatory structures aimed at protecting us from the putative risks of rDNA technology were enacted in the 1970s and 1980s, based on the knowledge and experience available at that time, but driven by a large amount of precautionary anxiety. Since then, there has been some regulatory 'fine tuning', but the biosafety regulations remain fundamentally as they were, and are in some cases stricter and/or more extensive. Even at the outset, most of the pioneering academic scientists developing genetically modified crops and foods found the new regulations excessive and ill-formed, as they focused on the process of GM, rather than the final product. That is, the academic scientific community determined very early that hazard is invariably associated with the final product, and that the breeding method is immaterial to hazard. Particularly inconsistent is the intensity of regulatory oversight of biotechnology compared to the lack of regulatory oversight over other products and breeding techniques such as ionizing radiation and other types of induced mutagenesis. Still, policymakers and the regulatory agencies have attempted to justify the original regulatory approach, arguing precaution and non-familiarity. That is, with a scant history of safe use of the technologies at the time, it was 'unfamiliar territory'; no one knew what unexpected or unintended hazards might arise, hence the perceived need to be extra careful. Today, after substantial experience with, and use of, rDNA products and biosafety regulations, is it time to revisit the 'non-familiarity argument' and biosafety regulations? Can safety regulations covering new technologies learn from the mistakes made in regulating agri-food biotechnologies?

The current safety regulations covering agri-food biotechnologies probably do not serve as a suitable template on which to build regulatory structures governing newer technologies, especially those involving new or modified processes and methods. Although the safety record of first and second generation agricultural biotechnologies to date is unblemished, that is less likely an achievement of the regulatory structures than attributable to the inherent safety of the technologies themselves. If anything, one might argue that the agbiotech regulatory milieu has actually increased risks to the populace and the environment due to the unnecessarily obstructive regulations without any compensatory increase in safety assurance. If we wish to learn from the mistakes of agri-food biotechnology regulations, we should design safety regulations based on the actual scientifically documented risks posed by new products, with a trigger for regulatory scrutiny based on the product not the process by which it is made, and we should impose scrutiny commensurate with the degree of risk actually presented, rather than on the risk perceptions.

Introduction

The first rDNA technologies were developed in 1973, with the successful fusion of fragments of DNA from different biological sources, spawning a rapid development and expansion of rDNA technological methods in many areas of scientific endeavour, as well as commercial investment in various applied aspects of the technology. From the beginning, the scientific community itself recognised biotechnologies could pose risks, leading to a short self-imposed moratorium while the potential hazards could be investigated and safety regulatory systems setup (Berg *et al.* 1974; NIH 1976; Berg and Singer 1995). Unfortunately, the fledgling regulatory safety structures were unable to keep pace with the early technological advances and had to be created hastily in response to perceived public anxieties instead of more solid scientific footings.

Problems with the biosafety regulations were apparent to scientists from the beginning. Because each jurisdiction developed their own regulations, there were no international standards or coordination. Indeed, some countries established multiple, different, and sometimes conflicting regulatory structures within their own jurisdictions (see McHughen 2000; 2007).

Such problems continue today. For example, there is no agreed standard definition of what is to be regulated. Although many use the terms 'GMO', 'GE', 'rDNA', and 'biotechnology' interchangeably, the regulations capture different things in different jurisdictions. This inconsistency has caused considerable political and economic problems due to disruptions to international trade. For instance, an importing country may receive shipment of a regulated product that is exempt in the exporting country. While international efforts like the Cartagena Protocol have attempted to alleviate this friction, it is important to note that while 161 countries have endorsed the Protocol, none of the major food commodity exporting countries, including the USA, Canada, Australia, and Argentina, are signatories to the Protocol (see Secretariat of the Convention on Biological Diversity 20 June 2011). Also, there have already been successful challenges to the World Trade Organisation (WTO) for erecting unscientific barriers to trade, such as the prominent case pitting USA, Canada, and Argentina against the EU for refusing shipments of biotech commodities without scientific evidence of hazard (WTO 24 February 2010), as is required under WTO rules, and as agreed to by the EU as a condition of membership.

In addition to the political and commercial frictions, the scientific community argues against the practice of employing the 'all-or-nothing' principle when a biosafety regulatory action is triggered. To illustrate, consider a common and popular food product like a tomato. A new variety of tomato – if it triggers regulatory action under one country's regulatory scheme – may be considered by the public to be such a massive threat to human health or the environment in that country that it requires a multimillion dollar investigation, with a subsequent clean bill of health and environmental safety confirmation, before it can be released to the market. However, that same tomato can be placed on the market in a neighbouring country with no safety oversight whatsoever, if the wording of the neighbouring country's regulations fails to capture the new tomato. In the professional scientific and medical communities, the tomato does not change its actual safety features in crossing the border – if it presents a hazard in one country, it will present the same hazard next door. A common maxim of true science based regulatory practice is that products should be scrutinised commensurate with the level of risk posed. A potentially highly hazardous tomato deserves a high degree of safety scrutiny; a low risk tomato should receive a modest amount of safety testing. In any case, the level of risk (and therefore the level of appropriate scrutiny) does not change going across a political boundary. Unfortunately, this sensible maxim is not a factor underpinning most agri-food biotechnology regulations.

Another problem is in the political manipulation of the regulatory language. In the EU, the regulations covering GM foods and crops capture those products made 'with' GM material, but exempt those made 'from' GM materials. A specific illustration is hard cheese made 'from' GM microbial chymosin (which, incidentally, is also advertised as 'suitable for vegetarians' because the standard curdling agent, rennet, is traditionally extracted from animal innards). In contrast, corn oil made 'with' GM corn is captured for regulatory review, even though the respective products lack transgenes or new transgene proteins and are in essence (as well as in degree of risk) identical to their traditional, non-GM counterpart foods. There is no scientific test to determine whether corn oil came from a GM corn plant or a non-GM corn plant. Nor is there any scientific evidence to suggest that the corn oil from GM corn plants presents any greater (or lesser, for that matter) risk to humans, animals, or the environment.

Such subtle regulatory distinctions, even if not lost on ordinary users of the English language, make no sense to the scientific community. In the political realm, however, such distinctions are simple: the products made 'with' GM ingredients, and therefore captured for regulatory scrutiny, are mainly imported from overseas markets, while products made 'from' GM material, and thus exempt, are primarily domestic, so the onerous regulations are preferentially directed against imports and protective of domestic markets (WTO 24 February 2010).

Furthermore, the problematic semantics continue with the non-scientific focus on 'process' rather than 'product'. In most countries, the biosafety regulations are triggered to capture products developed using proscribed processes of biotechnology. But in the scientific and medical communities, any hazard lies within the product, regardless of the process used to generate the products, so it is not scientifically valid to capture things based on the method of development; they should be captured according to the (potentially hazardous) features of the product itself. Every scientific panel to investigate this issue, including such prestigious groups as the US National Academy of Sciences (e.g. NAS 1987, 2000, 2004) and the Organisation for Economic Co-operation and Development (OECD 1986), has consistently upheld the concept that hazard is in the product, and so biosafety regulations triggered by process instead of product are not scientifically valid. Yet both the EU and the USA continue to use the scientifically baseless process trigger.

One lethal problem with any process-based regulatory trigger is that it assumes the proscribed process is static, when in actuality all functional technologies are dynamic. So it fails to capture modifications to the technology developed as a normal result of incremental innovation and improvement, where the intent is to improve the methodology gradually. Because the 'process' definition in regulatory language is written, it cannot anticipate the technical changes of the future. Another problem is that the process-based language fails to capture those innovations motivated solely or largely to circumvent the regulations. There are several examples in the area of agricultural biotechnology alone.

Cisgenics is the use of rDNA to clone and transfer genetic material from one organism to another of the same species (Cisgenesis 31 May 2011). Technically, this circumvents those definitions capturing 'trans' species transfer, and is being touted by some developers as a means to avoid regulatory scrutiny. Oligonucleotide-mediated mutagenesis (Breyer *et al.* 2009) is another technique with the potential to circumvent regulatory obstacles by carefully navigating the semantics of the regulatory capture language. Similarly, the increasingly popular RNAi technology, used to inactivate a gene, need not employ any foreign DNA (see e.g. Mohan Jain and Brar 2010), yet may or may not be captured by process-based regulatory schemes. Grafts are common in horticulture, where a rootstock of one species serves as the structural foundation of a scion from a different species. The fruit, arising on the scion, is not itself GM and required no rDNA technology, but the plant as a whole combines two different species. Such plants are technically captured as GM according to regulatory language of some jurisdictions, including the EU (Morris and Spillane 2008).

Selection of non-GM F2 progeny provides us with another example of the pitfalls of process-based regulation. If a GM plant is used as a parent in a breeding programme, the breeder may select traits other than those associated with the transgene. If a new cultivar is developed from a progeny line lacking the transgene, is the new line considered transgenic for the purpose of regulatory scrutiny? There is no clear answer to this question in most jurisdictions. If, instead, the trigger for regulatory scrutiny was based on the nature of their traits, potentially hazardous items would be captured regardless of their method of breeding. Using product, not process, as the basis for safety assessment is scientifically sound and provides a far higher level of confidence in the regulatory system.

International trade implications

Another problem with many current regulatory schemes, which together lack a single standard GM definition, is in international trade. A product in one place may be considered non-GM or exempt, but is captured when shipped into another jurisdiction, potentially resulting in trade disruptions. In Canada, for example, the so-called Clearfield™ crops (cultivars resistant to certain herbicides, but developed using somaclonal variation, or mutation breeding) are captured for regulatory review because they are considered 'Plants with Novel Traits' (PNTs), even though they are not products of rDNA technologies (see e.g. Canadian Food Inspection Agency 30 April 2003). In other countries, the Clearfield crops are exempt from such safety review. This is fine for Clearfield crops approved and grown in Canada, then exported overseas, but would be problematic if Clearfield crops not specifically approved in Canada were grown overseas – where they are exempt from biosafety regulatory scrutiny – and the commodity shipped to Canada, where the shipment would be subject to refusal due to lack of Canadian regulatory approval.

Scientific flaws in agbiotech regulatory scrutiny

In addition to issues with differential nomenclature, regulatory capture based on process of breeding fails to recognise that hazard, when present, is invariably associated with the final product, regardless of the process by which it was created. For example, cultivated rice is susceptible to a bacterial blight disease. Standard traditional breeding has not been able to overcome this disease, as there are no resistance genes available in cultivated rice varieties to serve as a source for crossing. However, there is a gene conferring resistance to bacterial blight, Xa21, in a rice relative. The Xa21 gene has been identified, characterised, isolated, and cloned using molecular techniques and transferred to cultivated rice varieties using rDNA techniques, resulting in a new rice variety with all the useful features of regular cultivated rice, but with the addition of blight resistance (Tu et al. 1998; Khush et al. 2001). As well, traditional rice breeders, including some of the same members as the biotech teams, and working with the same Xa21 gene, were able (through impressive, painstaking, and time consuming effort) to transfer the Xa21 gene into cultivated rice varieties using old fashioned crossing (Khush et al. 1990).

The end result was two lines of blight resistant cultivated rice; one from a traditional breeding process and one from a biotech process (Khush et al. 2001). From a safety perspective, both rice lines present equal degrees of risk, as they both carry the full complement of cultivated rice genes plus one additional feature, the Xa21 gene. If one line presents a certain risk, the other will present the same risk. If one is benign, the other is benign. However, in the current regulatory scheme, one line is assumed to be safe, whereas the other is assumed to be potentially hazardous, and requires strict and onerous regulatory scrutiny which may take years, and millions of dollars of safety assessment, before it can be released.

The purpose of agbiotech 'science based' regulations

A legitimate regulatory safety system is designed to protect human health and the environment from harms associated with new products. Certainly, any new product can carry some degree of risk, but that risk is not necessarily any greater than the products the new item will replace. A new cigarette lighter, whether using a heating coil or an open flame, will still carry some potential risks from igniting things other than cigarettes. It may also carry additional or heightened risks by being, for example, more susceptible to spontaneous ignition. So a risk assessment attempts to identify such potential risks before the product is released to the market. In the case of the lighter, the developer might be able to alter the design to make the lighter conform to safety standards, otherwise it may not be allowed on the market at all. Importantly, the identified hazard lies in the features of the final product – in this case a cigarette lighter – and not in the method used to manufacture the lighter.

In theory, the risk assessment procedure attempts to identify and then quantify any risks associated with the captured product. This first phase of the assessment process is a purely scientific endeavour conducted by technical experts familiar with the technology. The second phase of regulatory review is risk management, where the scientific information gathered in the risk assessment is reviewed by professionals who may be – but need not be – technical experts.[1] Their job is to determine how to mitigate any risks identified or, if sufficiently hazardous, whether the product should even be approved. If no additional or enhanced risks are identified, the product arguably should move toward commercialisation, unless there are legitimate non-scientific reasons to block access to the market. For example, as we have seen in some jurisdictions, ethical concerns over the use of embryonic stem cells may keep otherwise safe and efficacious products and services from commercial release. Alternatively, if identified risks can be mitigated, the risk managers might ask the developers to provide an appropriate mitigation, either directly to the product or to the means of usage by the end user. For example, in the case of an explosive cigarette lighter, the mitigation might involve a more robust fuel chamber to mitigate the likelihood of spontaneous ignition. In the case of genetically modified Bt crops, one mitigation strategy lies in the management of the cropping system, in which farmers are required to provide a refuge area for insects, so as to delay the onset of insect populations with resistance to the Bt product. That is, insects will sooner or later become resistant to Bt if they are constantly exposed to the Bt toxin. So farmers wishing to grow Bt crops must also grow non-Bt crops on a certain percentage of land nearby, as this refugia delays the onset of Bt resistant pests by reducing the constant exposure to Bt toxin in insect populations. Finally, some products might be so hazardous that no mitigation is suitable to reduce the risks to acceptable levels; such products are denied access to the market based on the documented and legitimate scientific grounds of excessive and unmanageable risk.

As mentioned at the beginning of this chapter, when transgenic technology was first developed in the early 1970s, scientists themselves recognised it was a powerful technology that might be hazardous and argued for a moratorium until the safety issues could be investigated (Berg *et al.* 1974; 1975). In the ensuing years, a number of scientific analyses aimed at studying the risks associated with rDNA technologies largely calmed the anxieties within the scientific and medical communities by establishing that risks posed by rDNA techniques were essentially the same as the risks posed by current technologies. That is not to say that rDNA was risk free, only that our regulatory safety bureaucracies were already familiar with the kinds of risks presented (Office of Science and Technology Policy 1986). Therefore, the scientific academies argued, there was no reason to construct new regulatory bureaucracies to handle rDNA, as we already have regulatory structure with expertise to provide the requisite safety assurances (NAS 1983, 1987, 2000, 2002, 2004).

However, this appeal from the various international scientific and safety regulatory communities fell on deaf ears in most jurisdictions, as special regulations and even regulatory bureaucracies were created solely to deal with the 'unknown' risks of GE. While the stated intention was to provide a greater degree of assurance to the populace that the risks of GE were being taken seriously, the apparent result was the opposite, with the populace expressing an increasing disquiet over the use of rDNA, especially in agriculture and food applications, which provided political fodder to escalate regulatory restrictions on agricultural products of biotechnology.

As mentioned earlier, the regulatory burden faced by agricultural products of biotechnology is so onerous as to serve as a deterrent to developers, who then turn to unregulated, older and perhaps riskier methods of plant breeding to circumvent the regulatory scrutiny. For example, some developers are touting oligomediated mutagenesis and 'cisgenics' as non-regulated methods of breeding, as they do not require the use of foreign gene transfer to introduce desired traits (Breyer *et al.* 2009). But mutagenesis is considered by the scientific community as a breeding method more likely than transgenesis to result in unexpected, potentially hazardous, products (NAS 2004). And cisgenics, the use of genes from related species as a source of traits for gene transfer, still requires rDNA to effect the insertion of the new trait. So, since the underlying assumption justifying the current safety regulations in most jurisdictions is that the process of rDNA is inherently a potentially hazardous activity, cisgenics is just as likely to generate a hazardous product as transgenics. Yet these developers argue that the regulations apply only to 'transgenic' rDNA, so 'cisgenic' rDNA is exempt.

Moreover, the same putatively hazardous and fearful rDNA technology is used to make most modern pharmaceuticals and many medical treatments, so any hazards inherent in the rDNA process would equally be present in these biotech products. But there is little or no public anxiety over these applications, and as with agbiotech applications, no history of problems in these products to support the underlying assumption. This contradiction in regulatory oversight between agricultural use of rDNA and medical rDNA has led some in the agbiotech community to wonder if the covert intent of the regulations was not to assure safety, but instead to hinder the use of biotechnology in agriculture. Whether or not this is so remains uncertain, but the onerous regulations certainly have presented a major obstacle impeding the commercialisation of new crops developed using rDNA as a breeding tool.

Currently, only a handful of biotech crops are cultivated worldwide. These are mainly soybean, corn, canola, and cotton (James 2011). Most of these crops were initially bred during the 1990s. Even though these crops have enjoyed remarkable success in terms of market share – for example, in capturing 93 per cent of the US soy and cotton acreage, and 86 per cent of the US corn acreages in 2010 alone (USDA 1 July 2010) – other examples of biotech crops are scant. Apart from virus resistant transgenic papaya (limited to Hawaii) and squash, and recent releases of herbicide tolerant sugar beet and alfalfa, the majority of the wide range of crops and traits developed and field tested so far remain uncommercialised (Graff *et al.* 2009; Miller and Bradford 2010). The USDA has authorised over 17,000 field trials of biotech plants since 1988, consisting of dozens of different species and a wide range of improved traits (Information Systems for Biotechnology 2010). However, in many cases, potentially valuable new cultivars have not been released because developers are unable to afford the costs and face the complexity of regulatory compliance (Kalaitzandonakes *et al.* 2007; Specialty Crop Regulatory Assistance 2011).

Placing the risks of agbiotech in context: how does the scientific community evaluate the hazards of rDNA crop and plant breeding?

The US National Academy of Sciences and the Institute of Medicine conducted a study into the safety of genetically engineered foods in 2002–04 (NAS 2004). As part of the assessment, they compared the likelihood of unexpected results as a proxy for potentially hazardous outcomes – as no one would intentionally try to create a hazardous food or crop – among several methods of plant breeding used by biotech and traditional plant breeders.

The chart in Figure 3.1 comparing these different methods of GM is telling, as it does not show 'GM' as being more hazardous, and 'traditional' as being more benign. Instead, the results show that the likelihood of unexpected results is a continuum, across which all forms of breeding, including traditional and modern, carry some degree of risk, but within which it is impossible to draw a line segregating GM methods from others. This means any attempt to regulate GM products apart from those created through other methods is scientifically invalid.

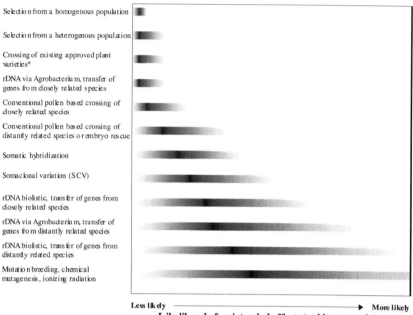

Selection from a homogenous population

Selection from a heterogenous population

Crossing of existing approved plant varieties*

rDNA via Agrobacterium, transfer of genes from closely related species

Conventional pollen based crossing of closely related species

Conventional pollen based crossing of distantly related species or embryo rescue

Somatic hybridization

Somaclonal variation (SCV)

rDNA biolistic, transfer of genes from closely related species

rDNA via Agrobacterium, transfer of genes from distantly related species

rDNA biolistic, transfer of genes from distantly related species

Mutation breeding, chemical mutagenesis, ionizing radiation

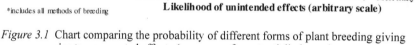

Less likely ⟶ More likely

*includes all methods of breeding   **Likelihood of unintended effects (arbitrary scale)**

Figure 3.1 Chart comparing the probability of different forms of plant breeding giving rise to unexpected effects (as a proxy for potentially hazardous outcomes). The graph illustrates the relative likelihood of an unexpected effect (with likelihood increasing the dark band to the right) and also the intensity of an unexpected effect, from mild to more severe (the density of the dark band).

Source: NAS (2004) Reprinted with permission from *Safety of Genetically Engineered Foods: Approaches to Assessing Unintended Health Effects* (2004) by the National Academy of Sciences, Courtesy of National Academies Press, Washington, D.C.

Unfortunately, most regulatory bureaucracies worldwide follow the flawed premise that products of rDNA are (somehow) inherently of higher risk than non-rDNA products. Proponents of this regulatory scheme might point to the outcomes – that there remain no documented cases of harm from approved biotech crops of foods – and suggest the safety record is due to the strict regulatory scrutiny ensuring the approved lines are indeed safe. But this is fallacious reasoning. If legitimate, the argument would be supported by exposing those hazardous candidate biotech crops failing the regulatory review and thus being denied market access. In fact, few candidate biotech plants and crops are shown to be hazardous even after undergoing strict regulatory safety review, and there is certainly no basis for categorical hazard, which is the basis for the regulatory bias capturing all biotech crops and only biotech crops.

The onerous regulatory approval process serves as a barrier to the release of additional but unknowable numbers of safe and beneficial biotech crops. However, this is not the only penalty for maintaining an unscientific regulatory scheme. Other penalties include the increase in risk carried by devoting scarce regulatory resources to known low risk products of biotech, thus diverting those scarce resources away from potentially higher risk products, such as foods with higher levels of naturally occurring toxins or allergens, introductions of invasive species from foreign countries, or other potentially hazardous products currently exempted from biosafety regulatory scrutiny (McHughen 2007). Eventually the odds will catch up and damage will occur elsewhere on the safety regulatory front, in an area not related to biotech at all.

How do we properly and safely manage products of agricultural biotechnology?

First, we need to recognise that agricultural biotechnology is here to stay. Bans do not work. Countries that attempt to ban a useful technology merely drive it underground, as happened in Brazil until the now near-ubiquitous GM soybeans were approved. There are real risks with some agbiotech products; those real risks need to be properly managed. For example, growing new insect resistant crops will eventually lead to insect pests with resistance to the crop, diminishing the protection offered by the new crop. Appropriate management demands strategies (such as refugia, discussed earlier) to delay the onset of pest populations with resistance so as to maximise the insect control of the crop. Importantly, such a management strategy is required whether or not the pest resistant crop was developed using biotechnology.

Unlike politics, actual risk does not fluctuate with the vagaries of popular fashion and public perception. Risk assessment when conducted properly is a scientific endeavour, not a political exercise. Such assessment should be based on scientifically validated procedures to identify actual hazards and the probability of those hazards coming to fruition. There should be no place in the risk assessment for intrusion of political interference. Socio-economics, marketplace issues or public anxieties, whether real or otherwise, only distort the scientific analysis of risk. This is not to say these non-scientific considerations are socially irrelevant, but they cannot be legitimately inserted into the risk assessment phase without jeopardizing the legitimacy of the regulatory safety system, leading to increased likelihood of harm.

Once a scientific risk assessment has laid a foundation by identifying and documenting the scientifically valid concerns with a given product, the risk management phase should then build policy atop that foundation. In valid risk management, scientists and others consider the kinds of risks presented and the management tools available to mitigate the identified risks. It may be that the product is so hazardous that no satisfactory mitigation tools are available, so restricting or precluding release is the appropriate management decision. Past experience with rDNA technologies shows, however, that risks can be appropriately managed, and the product should be allowed onto the market so society can benefit from it. The risk management phase is where non-scientific aspects can be presented to influence the ultimate fate of regulations and regulated products. For example, society may have ethical issues with the destruction of embryonic stem cells; risk management is the phase where societal abhorrence is evaluated and the product may be restricted, even if the product has no particular safety issues per se. People may hold strong beliefs that something is hazardous, but regardless of the intensity or strength of the belief, or the number of people holding the belief, it does not change the actual scientific hazard status or create a hazard where none actually exists.

Conclusions

Agricultural biotechnology is a set of breeding methods. The products resulting from the use of those methods pose risks similar to products developed using other, traditional, breeding methods. All arguments to regulate rDNA apply to conventional breeding as well. (Why are we not regulating those?)

The current regulatory governance of agricultural biotechnology suffers several fundamental flaws. One flaw is the failure to recognise that hazard is carried by the product, not by the process of development. This flawed attitude led to an exclusive regulatory focus on the process of breeding (i.e. rDNA) rather than the resulting products. Another flawed principle holds that agri-food biotech regulatory scrutiny is case-by-case and 'all-or-nothing', meaning that every GM plant variety is subject to immense degree of regulatory oversight, while products of non-rDNA technologies posing similar risks face no safety scrutiny whatsoever. One illustration of this point should suffice: both American and European consumers are permitted to consume more arsenic, a known toxin, than they are unapproved GM foods (Smith *et al.* 2002). The regulatory maxim that scrutiny should be commensurate with degree of risk was ignored with agri-food biotechnologies.

For the purposes of developing next generational agbiotech regulatory regimes, we should learn from our past experience that a legitimate biosafety regulatory system is truly science based; that the trigger for scrutiny should be the features of the product regardless of the process by which the product was made; and that the degree of scrutiny should be commensurate with the degree of risk posed.

Note

1 This two-phase risk assessment followed by risk management is standard practice worldwide – at least in theory. Risk managers are usually professionals, but need not be technical experts if they have appropriate technical experts working in close collaboration.

Bibliography

Berg, P. and Singer, M.F. (1995) 'The recombinant DNA controversy: Twenty years later', *Proc. Natl. Acad. Sci.* Vol 92,: 9011–3.

Berg, P., Baltimore, D., Boyer, H.W., Cohen, S.N., Davis, R.W., Hogness, D.S., Nathans, D., Roblin, R., Watson, J.D., Weissman, S. and Zinder, N.D. (1974) 'Biohazards of recombinant DNA', *Science* 185: 303.

Berg, P., Baltimore, D., Brenner, S., Roblin, R.O. III and Singer, M.F. (1975) 'Summary statement of the Asilomar Conference on recombinant DNA molecules', *Proceedings of the National Academy of Science*, USA 72, 1981–1984. (also in *Science* 188: 991).

Breyer, D. Herman, P., Brandenburger, A., Gheysen, G., Remaut, E., Soumillion, P., Van Doorsselaere, J., Custers, R., Pauwels, K., Sneyers, M. and Reheul, D. (2009) 'Genetic modification through oligonucleotide-mediated mutagenesis. A GMO regulatory challenge?', *Environmental Biosafety Research*: 1–8.

Canadian Food Inspection Agency, 'Decision Document DD2003-44 Determination of the Safety of BASF's Imazamox Tolerant (CLEARFIELD™) Wheat AP602CL' (last modified 30 April 2003) http://www.inspection.gc.ca/english/plaveg/bio/dd/dd0344e. shtml, accessed 31 May 2011.

Cisgenesis, White Paper: 'Faster and Cheaper Approval Needed for Safe Genetically Modified Crops Without Foreign Genes (Cisgenic Crops)', http://www.cisgenesis.com/ content/view/4/28/lang,english/, accessed 31 May 2011.

Graff G.D., Zilberman, D. and Bennett, A.B. (2009) 'The contraction of agbiotech product quality innovation', *Nature Biotechnology* 27: 702–4.

Information Systems for Biotechnology (2010) 'USDA Field Tests of GM Crops' http:// www.isb.vt.edu/search-release-data.aspx, accessed 31 May 2011.

James, C. (2011) 'Global status of commercialised biotech/GM crops: 2010', *ISAAA brief* 42 (http://www.isaaa.org/). Access date: 25/04/2011.

Kalaitzandonakes, N., Alston, J.M. and Bradford, K.J. (2007) 'Compliance costs for regulatory approval of new biotech crops', *Nature Biotechnology* 25: 509–11.

Khush, G.S., Bacalangco, E. and Ogawa, T. (1990) 'A new gene for resistance to bacterial blight from *O. longistaminata*'. *Rice Genetics Newsletter* 7: 121–2.

Khush, G.S., Brar, D.S. and Hardy, B. (2001) *Rice Genetics IV*. IRRI. Enfield, New Hampshire: Science Publishers, Inc.

McHughen, A. (2000) *Pandora's Picnic Basket: The potential and hazards of genetically modified foods*. Oxford and New York: Oxford University Press.

McHughen, A. (2007) 'Fatal flaws in agbiotech regulatory policies', *Nature Biotechnology* 25:, 725–7.

Miller, J.K. and Bradford, K.J. (2010) 'The regulatory bottleneck for biotech specialty crops', *Nature Biotechnology* 28: 1012–4.

Mohan Jain, S. and Brar, D.S. (2010) *Molecular Techniques in Crop Improvement*, 2nd edition. New York: Springer.

Morris, S. and Spillane, C. (2008) 'GM directive deficiencies in the European Union', *EMBO Reports*, 9:1–5.

National Academy of Sciences (1983) *Risk Assessment in the Federal Government. Managing the Process*. Washington DC: National Academies Press.

National Academy of Sciences (1987) *Introduction of Recombinant DNA-Engineered Organisms into the Environment: Key Issues*. Washington DC: National Academies Press.

National Academy of Sciences (2000) *Genetically modified pest protected plants: Science and regulation*. Washington DC: National Academies Press.

National Academy of Sciences (2002) *Environmental Effects of Transgenic Plants*. Washington DC: National Academies Press.

National Academy of Sciences (2004) *Safety of Genetically Engineered Foods*. Washington DC: National Academies Press.

NIH (1976) 'NIH Guidelines for Research Involving Recombinant DNA Molecules', *Federal Register* 41 June: 27911.

OECD (1986) *Recombinant DNA Safety Considerations: Safety Considerations for Industrial, Agricultural and Environmental Applications of Organisms Derived by Recombinant DNA Techniques*. OECD: Paris.

Office of Science and Technology Policy (1986) 'Coordinated framework for regulation of biotechnology: announcement of policy and notice for public comment', *Federal Register* 51: 23302–9.

Secretariat of the Convention on Biological Diversity, 'The Cartagena Protocol on Biosafety to the Convention on Biological Diversity' (updated 20 June 2011) http://bch.cbd.int/protocol/, accessed 2 May 2011.

Smith, A.H., Lopipero, P.A., Bates, M.N. and Steinmaus, C.M. (2002) 'Arsenic Epidemiology and Drinking Water Standards', *Science* 296: 2145–6.

Specialty Crop Regulatory Assistance (2011) http://www.specialtycropassistance.org/, accessed 31 May 2011.

Tu, J., Ona, I., Zhang, Q., Mew, T.W., Khush, G.S. and Datta, S.K. (1998) 'Transgenic rice variety "IR72" with Xa21 is resistant to bacterial blight', *Theoretical Applied Genetics* 97: 31–6.

United States Department of Agriculture Economic Research Service, 'Data Sets: Adoption of Genetically Engineered Crops in the US' (updated 1 July 2010) http://www.ers.usda.gov/Data/BiotechCrops/, accessed 31 May 2011.

World Trade Organization, Dispute Settlement DS291, 'European Communities – Measures Affecting the Approval and Marketing of Biotech Products' (Updated 24 February 2010) http://www.wto.org/english/tratop_e/dispu_e/cases_e/ds291_e.htm, accessed 31 May 2011.

Part 2

Regulatory regime development theory and practice

4 Regulatory lifecycles and comparative biotechnology regulation

Analysing regulatory regimes in space and time

Michael Howlett and Andrea Migone

Introduction: understanding the spatial and temporal dimensions of national regulatory regimes

Understanding the nature and origins of national regulatory regimes is of interest to both proponents and opponents of new biotechnologies as well as to students of other sectors facing similar regulatory challenges from scientific and technologically innovative activity – such as nanotechnology, genetically modified organisms (GMOs), or synthetic biology (Bowman and Hodge 2007; Hodge *et al.* 2007; Furger *et al.* 2007; Kuzma and Tanji 2010). As the OECD has noted, 'regulation and the predictability of the regulatory environment influence the direction of biotechnology research, the types of research that are commercially viable, and the costs of research and development' (OECD 2009:144). Hence, understanding the range of possible variations in regulatory regimes and the factors driving their evolution and development is crucial in understanding the processes of discovery, innovation, and commercialisation in these fields.

Two general aspects of any regulatory regime must be investigated when dealing with specific cases. The first is spatial, as regulatory regimes formulated at the domestic level are often linked to transnational or international developments, while national regimes often exist as 'local variations' of more generic regulatory models. The second is temporal, since regulatory regimes change and evolve over time. Understanding how a regime emerges and changes is a necessary part of understanding the evolution of its 'local variations'. This chapter explores both of these spatio–temporal dimensions in the case of agri-food genomics regulation and constructs a general model of a regulatory regime and of a regulatory regime lifecycle. Drawing on comparative analysis of existing biotechnology regimes, we find that the spatial dimension is dominated by the nature of the approach taken by authorities towards controversial technologies – ranging from promotional to preventative with several steps in between – as well as the nature of the policy development process, whether led by state or societal actors. On the temporal dimension, we examine several overviews of regulatory development processes in several countries and sectors. We find that sectors follow the same steps in a generalised pattern of regulatory regime development, but that the detailed expression of this pattern differs by country and sector in terms of the positioning of an issue like biotechnology or genomics in the stages of regime development. Thinking in terms of these two dimensions allows us to analyse more precisely local variations in national regimes in terms of the comparative nature of the regime, its level of maturity, and the impact these characteristics have on activities such as product commercialisation and innovation.

Regulation, regulatory regimes and regulatory lifecycles

The use of the coercive power of the state to achieve government goals through the control or alteration of societal (and governmental) behaviour is the essence of many common types of governing instruments, especially regulation (Hood 1986). A good general definition of regulation is 'a process or activity in which government requires or proscribes certain activities or behaviour on the part of individuals and institutions, mostly private but sometimes public, and does so through a continuing administrative process, generally through specially designated regulatory agencies' (Reagan 1987: 15). Here regulation is seen as a governmental prescription, which must be complied with by intended targets on pain of penalties that range from the financial to criminal. While some regulations are *laws* being enforced by the police and judicial system, most are *administrative edicts* created under the terms of enabling legislation and administered on a continuing basis by a government department or specialised, quasi-judicial government agencies (Rosenbloom 2007). This is often referred to as 'command and control' regulation since its essence is to have a government issuing a 'command' to a target group in order to 'control' its behaviour. 'Control' also sometimes refers to the need for governments to continually monitor and enforce target group activity in order for a 'command' to be effective in the long term.

'Command and control' regulation is very commonly used for normative purposes and can include criminal laws, common laws, and civil codes (Cismaru and Lavack 2007; May 2002) but is also found in many areas of economic regulation of established markets for goods and service production. However, if regulation is resisted by target companies and industries (Baldwin and Cave 1999; Crew and Parker 2006; Scholz 1991), or if governments do not have the desire, capacity, or legitimacy required to enforce their orders, then several other regulatory methods and arrangements can be used where rules are more vague and the threat of penalties may be, at best, remote. These include the development of voluntary standards or various forms of delegated regulation in which target groups police their own activities. These different possible arrangements, which form the basis of any regulatory regime, include direct government regulation, the use of independent regulatory commissions (Majone 2005; Jacobzone 2005; Stern 1997; Gilardi 2005), delegated (Brockman 1998; Kuhlmann and Allsop 2008; Elgie 2006; Sinclair 1997; Tuohy and Wolfson 1978; Trebilcock *et al*. 1979), and voluntary regulation[1] (Elliott and Schlaepfer 2001; Schlager 1999; Cashore *et al*. 2003; Borraz 2007).

Different national regulatory regimes are composed of these different kinds of regulatory arrangements, sometimes in mixes but often in relatively pure form, which evolve over time. The result is a considerable range of variation in regulatory regimes in terms of both national preferences and sectoral patterns. These variations can be quite significant, extending from the manner in which regulations are enforced to alterations in the basic type of regulation used in a specific sector or issue area.

This begs the question of whether or not there is any general pattern of regulatory regime formation and evolution or if the arrangements found in specific sectors, such as biotechnology, are simply the result of more or less random or contingent sets of occurrences. Most observers argue the former (de Vries 2010), leading to different attempts being made in the literature on the subject to develop models to describe and explain the formation and dynamics of regulatory agencies and their activities.

One of the oldest and time-tested conceptions of regulatory evolution is that put forward by Marver Bernstein in his 1955 analysis of the development and spread of the regulatory institutions in the United States following the creation of the Interstate Commerce Commission in 1887. Bernstein postulated a basic genealogical or lifecycle model in which an agency moves from 'birth', when action to address a problem is first taken; through 'adolescence', when it grapples with ongoing issues and actors in the attempt to 'find its feet'; then to a long stage of 'maturity' in which its actions are heavily routinised; finally entering a period of 'decline' when it fails to keep up with changes in its environment, which ultimately ends in 'death' once its original purpose has passed (Bernstein 1955).

In this model, it is expected that initial regulatory arrangements will undergo many changes between 'birth' and 'maturity', after which they will be more or less locked in. While Bernstein did not address exactly how these changes occur, observation of the evolution of other regulatory issue areas – such as chemical toxic regulations, genomics, nanotechnology, and other sectors – has highlighted some of the processes present at the early stages of the regulatory lifecycle (Boucher 2008; Seaton *et al.* 2009). Experiences with de-regulation in the 1990s, in areas such as transportation and utilities, have highlighted some of the relevant forces at work in the later stages of regulatory regime decline.

Based on this work, Otway and Ravetz (1984) proposed a three stage model of early regulatory regime development – those most germane to 'new' and 'emerging' technology sectors such as biotechnology or genomics – proceeding from the recognition of a hazard, through the development of some limit values or standards for it, and finally to their implementation (see Table 4.1). In this 'linear model' specific kinds of regulatory activity are associated with each phase in a standard-building process, from collecting data to monitoring hazard occurrence and, finally, to the preparation of codes, implementation of inspections, and enforcement.

Table 4.1 The linear model after Otway and Ravetz (1984)

Stage	Issue	Regulatory Activity
Scientific Phase	Recognise existence of hazard	Collect medical/ecological information
Technical Phase	Define practical limit values for hazard producers	Monitor process/facility for hazards
Administrative Phase	Implementation	Prepare operating and inspection codes; implement inspections, advice, sanctions, etc. Iterate as experience is gained

Source: Otway and Ravetz (1984)

However, the model says little about the specific activities taking place in the early technical stage of standard development that Leiss (2001) and others have argued is a critical and often time-consuming stage. In his own work Leiss (2001), for example, also follows a linear, three stage logic but is clearer in dividing Otway and Ravetz' 'technical' phase into the two substages (see Table 4.2) falling between the very early 'recognition' or 'Scientific' phase (which Leiss assumes) and the later mature 'implementation' or 'Administrative' phase.

The combined Leiss-Otway-Ravetz model is contained in Table 4.3 and can be seen as an effort to fill in the missing gaps in Bernstein's 'adolescent' stage of development: to continue the metaphor, adding a stage of 'infancy' that is missing in Bernstein's original formulation.

Investigating the development of nascent Bio- and Nanotechnology-regulatory regimes, Howlett and Migone (2010) expanded on this developmental or genealogical logic by developing a multi-staged process of administrative

Table 4.2 Stages of risk controversy after Leiss (2001)

Stage	Characteristics	Problem	Regulatory Response
Early (10–15 years)	Incomplete hazard specification; Poor exposure assessment; Issue characterization	Scientific uncertainty; Lack of clinical trials; Stigmatization	Downplay scope of hazard; Let sleeping dogs lie; Denial
Middle Stage (5–10 years)	Science underway; Early epidemiology	Issue debates; Internationalization	Spin; Venue shifting
Mature Stage (decades)	Better science and epidemiology; Issue capture	Stakeholder/Media links; Popular frame; Agreements	Bilateral negotiations; Routinisation

Source: Leiss (2001)

Table 4.3 Stages of regulatory development after Bernstein (1955), Otway and Ravetz (1984) and Leiss (2001)

Bernstein Stage (modified)	Otway and Ravetz Phase	Leiss Stage	Characteristics	Problem	Regulatory Response
Birth	Scientific Phase		Recognise existence of hazard		
Infancy	Technical Phase	Early (10–15 years)	Incomplete hazard specification; Poor exposure assessment; Issue characterisation	Scientific uncertainty; Lack of clinical trials; Stigmatization	Downplay scope of hazard; Let sleeping dogs lie; Denial
Adolescence	Technical Phase	Middle Stage (5–10 years)	Science underway; Early epidemiology	Issue debates; International-isation	Spin; Venue shifting
Mature	Administrative Phase	Mature Stage (decades)	Betters cience and epidemiology; Issue capture	Stakeholder/ Media links; Popular frame agreements	Bilateral negotiations; Routinisation

Source: Bernstein, Otway and Ravetz (1984) and Leiss (2001)

adaptation to uncertainty in standard formation (see Table 4.4) beginning with the birth of a regime in crisis and controversy over public protection.

This model counted as many as four stages before an agency or rule enters into adolescence. Following the birth of the regime in a 'Pre-regulatory' stage of problem recognition, where a 'problem' emerges as accidents, state, or societal pressure brings the new issue area onto the regulatory agenda (Fleischer 2010) – what Otway and Ravetz termed the 'Scientific' stage – there is often a period of 'Adaptive Experimentation'. In this, an attempt is made to adapt existing statutes and rules to current problems, using existing regulatory tools and agencies to attempt to cover emerging issue areas. If this fails to address the problem, this period is followed by a period of 'Standard Seeking' when there is a desire to create new rules and standards but no clear knowledge of what they should be (Majone 2010; Borraz 2007). Both Otway and Ravetz described this as a key part of the 'Technical' or 'early' and 'middle' stages, labelled as the 'infant' stage in Table 4.3. A key element during this infant or juvenile stage is dealing with uncertainty and the fear of possible over or under regulation (Nichols and Zeckhauser 1988; Mashaw 1988; Aizenman 2009).

Table 4.4 Early stage regulatory lifecycle model

Phase	Characteristics	Activities
Pre-regulatory	Emergence of a 'problem' on the agenda	Accidents, state pressure, and societal pressure bring the new issue area on the agenda
Adaptive Experimentation	Attempt to adapt existing statutes and rules to current problems	Use of existing regulatory tools and agencies to cover emerging issue areas
Standard-Seeking	Desire to create new rules but no clear knowledge of what these rules/standards should be	Leads to various efforts to find the appropriate standard, including: a. Funding research b. Using 'principles' rather than standards c. Trying to get the problematisers (e.g. industry or science, as the case may be) to come up with the standards through various forms of 'voluntary' activity
Soft Regulation	Emergence of a 'light' hierarchical hand or 'soft' regulation	Involves the use of third-party certifications, more voluntary schemes.
Direct State Regulation	Emergence of more 'direct' state regulation	Results in the creation of either an Independent Regulatory Commission or Direct Administrative Rule-Making. Often this happens when scandals erupt, accidents occur, or complaints arise, then the level of state authority and presence is ramped up

Source: Howlett and Migone (2010)

If we blend these three models with Bernstein's idea of a regulatory lifecycle, we can generate a more nuanced model of the substages of regulatory evolution having nine stages in total. In this model, it is expected that initial regulatory arrangements will undergo a large number of changes between 'birth' and 'maturity' as the agency moves through periods of 'infancy', 'childhood', and 'adolescence', after which they will be more or less locked in to the standards and processes developed in the pre-adult stages. Five distinct periods are identifiable between the immediate birth of a regime and its 'mature' phase of development (see Table 4.5).

Table 4.5 Nine-stage model of regulatory regime lifecycle

Bernstein Lifecycle Stage	Phase	Issues	Task	Administrative Techniques
I. Birth	Pre-regulatory	Emergence of a 'problem' on the agenda	Dealing with accidents, state pressure, and societal pressure brings the new issue area on the agenda	Creation or designation of primary regulatory agency
II. Infancy	Co-optive	Attempt to adapt existing statutes and rules to current problems	Efforts at issue suppression	Delay or use of existing regulatory tools and agencies to cover emerging issue areas
III. Early Childhood	Early Scientific Phase (1–2 years)	Successful issue re-framing/stigmatisation	Poor knowledge base, heavily symbolic/ discursive struggles	Institution staffing, etc. Trying to get the problematisers (e.g. industry or science, as the case may be) to come up with the standards through various forms of 'voluntary' activity
IV. Late Childhood	Middle Technical Phase (8–10 years)	Desire to create new rules but no clear knowledge of what these rules/standards should be due to incomplete hazard characterisation and poor exposure assessment	Standard-Seeking	Adaptive Experimentation
			Large-scale research programs for hazard characterisation	Using 'principles' rather than standards
		Lobbying and back-room deal-making/lobbying.	Initial quantitative risk assessments	Emergence of a 'light' hierarchical hand or 'soft' regulation
		Venue hopping		

(Continued)

Table 4.5 (Continued)

Bernstein Lifecycle Stage	Phase	Issues	Task	Administrative Techniques
V. Adolesence	Late Administrative Phase (1–2 years)	Completion of hazards assessment Development of standards Frozen issue frames Issue ownership by specific groups	Smaller-scale maintenance research and legal/judicial activities	Emergence of more 'direct' state regulation Often this happens when scandals erupt, accidents occur, or complaints arise, then the level of state authority and presence is ramped up
VI. Young Adulthood	Precedent Building	Legal actions	Court activities	
VII. Maturation	Routinisation	Normalisation of the regulatory issues	Administrative activity	Emergence of specific agencies that 'own' the regulatory area
VIII. Decrepitude	Capture	Regulatory capture, Emergence of clientelism, Issue capture by institutionalised groups	Maintaining a favourable environment for the regulatees	Self-Regulation
IX. Death	Termination/ Reform	Modification/Death of the issue		De/Re-regulation

Sources: Bernstein, 1955; Otway and Ravetz 1984; Leiss, 2001; Hood *et al.* 1999a

The agri-food biotechnology/genomics case

Given the kinds of possible variations in regulatory regime configurations listed above, it should not be surprising that each regulatory regime is somewhat idiosyncratic. Different countries can adopt similar or different types of regulation, or elements within different types, in their regulatory arrangements and this can easily result in different countries exhibiting distinct differences in regulatory style, process, and content (Knill 1998). And different countries do often feature different propensities and tendencies to use particular types of regulatory tools or arrangements.

However, there is also a distinct tendency for regimes to exhibit fewer differences than would be expected, leading to notions of 'regulatory convergence' (Vick 2006; Holzinger, Knill and Sommerer 2008) or the idea that regulatory regimes in different countries, although they may begin in an idiosyncratic fashion, draw closer together and become more similar over time. That is, on a temporal level, different countries may proceed to regulation at different times and in different ways, but eventually emerge as 'lagged copies' of each other, in which some countries emerge as regulatory leaders and others as laggards or emulators. This next section provides an analysis of the case for agri-food biotechnology and genomics regulation.

Understanding local variation in national biotechnology regulatory regimes: the spatial dimension

Agri-food and biotechnology actors in general deal with seven distinct activities: intellectual property rights; public information inclusiveness of deliberation; retail and trade issues; safety and human health; consumer choice; public research investment; and commercialisation of biotechnology-related products (Haga and Willard 2006). These seven areas of activity cross-over into five areas of regulatory concern related to the monitoring and control of research activities; the development and implementation of legal frameworks; dealing with a variety of economic aspects of biotechnological innovation; as well as various education and implementation issues (Haga and Willard 2006: 967). A wide range of specific regulatory problems/issues thus exists where these seven areas of activity and five areas of regulatory concern intersect. For example, at the intersection between public information/deliberation activities and regulatory economic concerns we find issues such as who bears the costs of consultation. Using this lens, biotechnology regulatory policy-making in general, and agri-food genomics regulation in particular, can be seen as involving the design and adoption of a set of policies to deal with the presence or absence of the previously-noted regulatory issues in the context of specific technologies and local circumstances (Montpetit 2005; Sheingate 2006) (see Appendix 1).

While each technology and jurisdiction thus forms a distinct regulatory 'space' (Paarlberg, 2000) we can use earlier work to simplify the search for commonalities in different countries' approaches. Paarlberg, for example, compresses Haga and Willard's seven categories of biotechnology into five, combining public information and consumer activities and those related to commercialisation and research. National regulatory responses can then be scored in terms of their general orientation towards the promotion or discouragement of the biotechnology in question. This generates a country-specific measure of the overall regulatory approach in terms of its orientation towards the biotechnology in question: they range from 'promotional', 'permissive', or 'precautionary' to a 'preventive' response. Regulations that accelerated the spread of GM crop and food technologies within the borders of a nation, for example, can be termed 'promotional'. Those more neutral toward the new technology, in tending neither to speed nor to slow its spread, are 'permissive'. Regulations intended to slow the spread of GM crops and foods are 'precautionary', while those that tended to block or ban entirely the spread of technology are 'preventive' (Paarlberg 2000: 4) (see Table 4.6).

Biotechnology regulatory issues have been tackled with one, or a mix, of several basic regulatory approaches (Haga and Willard 2006: 968). These range from public consultations to a more legislative approach, and include voluntary approaches, guidelines, and regulation, as well as some instances of stricter state control including legislative, regulatory, and guideline approaches. Combining the latter into a single 'top-down' category and the former into a 'bottom-up' one, allows us to link these approaches to Paarlberg's categories of regulatory orientations and create the comparative national regulatory matrix shown in Table 4.7 below.

We can locate different countries and sectors of activity within four policy quadrants reflecting the country's preferences through its dominant orientation and direction towards specific biotechnology developments. Two types appear in US and EU biotechnology regulation. Both spring from the concerns embodied in the Berg Letter (see below) (Berg *et al.* 1974) but rapidly part company. The US developed a promotional approach based on the substantial equivalency principle

Table 4.6 Paarlberg model of policy options and regimes towards GM crops

	Promotional	*Permissive*	*Precautionary*	*Preventive*
Intellectual property rights	Full patent protection, plus plant breeders' rights (PBR) under UPOV 1991	PBRs under UPOV 1991	PBRs under UPOV 1978, which preserves farmers' privileges	No IPRs for plants or animals or IPRs on paper that are not enforced
Biosafety	No careful screening, only token screening, or approval based on approvals in other countries	Case-by-case screening primarily for demonstrated risk, depending on intended use of product	Case-by-case screening also for scientific uncertainties owing to novelty of the GM process	No careful case-by-case screening; risk assumed because of GM process
Trade	GM crops promoted to lower commodity production costs and boost exports; no restrictions on imports of GM seeds or plant materials	GM crops neither promoted nor prevented; imports of GM commodities limited in the same way as non-GM in accordance with science-based WTO standards	Imports of GM seeds and materials screened or restrained separately and more tightly than non-GM; labelling requirements imposed on import of GM foods or commodities	GM seed and plant imports blocked; GM-free status maintained in hopes of capturing export market premiums

(Continued)

Table 4.6 (Continued)

	Promotional	Permissive	Precautionary	Preventive
Public information and consumer choice	No regulatory distinction drawn between GM and non-GM foods when either testing or labelling for food safety	Distinction made between GM and non-GM foods on some existing food labels but not so as to require segregation of market channels	Comprehensive positive labelling of all GM foods required and enforced with segregated market channels	GM food sales banned or warning labels that stigmatise GM foods as unsafe to consumers required
Public research investment and commercialisation	Treasury resources spent on both development and local adaptations of GM crop technologies	Treasury resources spent on local adaptations of GM crop technologies but not on development of new transgenes	No significant treasury resources spent on either GM crop research or adaptation; donors allowed to finance local adaptations of GM crops	Neither treasury nor donor funds spent on any adaptation nor development of GM crop technology

Table 4.7 Mapping comparative biotechnology regulatory regimes

Procedural Direction		Regulatory Orientation
State-driven/Top-Down	Public-driven/Bottom-up	Promotion
Type I – State-Promotional/Permissive e.g. US	Type II – Public-Promotional/Permissive e.g. Denmark	Permissive
Type III – State-Precautionary/Preventative e.g. Canada	Type IV – Public-Precautionary/Preventative e.g. Italy	Precautionary Preventive

Source: Howlett and Migone (2010)

and infused with a predominantly scientific logic. The EEC/EU (European Economic Community/European Union) instead developed a precautionary approach mixing scientific and social logics. Most countries' approaches exist in an interdependent relation with international standards and rules comprising OECD standards and protocols like the Codex Alimentarius or the Cartagena agreement on biodiversity. While these approaches show variations, on balance they have a more promotional feel and tilt towards soft regulation. More direct state regulation is found only infrequently in the biotechnology sector. For example, the European Union Directive 2001/18, which regulates the deliberate release of GMOs in the member states of the Union (Cantley 2007).[2]

This is a good model for describing how different national policy and regulatory orientations are similar and how they differ (see Howlett and Migone 2010). However, it provides only a single snapshot of an evolving situation. To more fully describe and understand regulatory regimes, a more fluid evolutionary or genealogical perspective is needed. A purely spatial analysis, while aiding the identification of the significant components of regulatory regimes, reveals neither which countries in each sector are leaders and which laggards, nor the extent to which each regime has evolved along the regulatory lifecycle.

Understanding the evolution of biotechnology regulation: the temporal dimension

In many countries biotechnology and especially genomics regulation remains at a 'late childhood stage' or has only recently moved towards 'adolescence' or 'young adulthood'. Even after a lengthy 'search for standards' period (Murphy and Levidow 2006), most regimes remain at the stage of 'soft regulation', if not earlier. Often simple additions to existing legislation that dealt with food safety and agricultural practices have been the only changes made to incorporate new innovations such as various kinds of GM products (Canada 2001; Howlett and Migone 2010). Existing institutions like the US Food and Drugs Administration and the Canadian Food Inspection Agency have been drafted to administer these areas in an 'adaptive experimentation' fashion.

In the mid-1970s, the scientific community began worrying about creating some signposts for emerging biotechnology. In 1974, the Berg Letter appeared in *Science* and *Nature*, drawing attention to the potential hazards posed by early innovations involving recombinant DNA; the following year, the Asilomar Conference kick-started a broader debate about biotechnologies. The next issue to emerge was the patentability of living organisms. In 1980, the US Supreme Court in a 5–4 decision allowed a genetically modified bacteria to be patented on the grounds that it did not occur naturally [Diamond v. Chakrabarty, 447 US 303 (1980)], setting precedent for the Harvard Mouse case of 1988. The early 1980s brought home the reality of recombinant DNA applications and generated a series of adaptive efforts in both the EU and the US, which began to establish themselves as the two most important and distinct poles for biotechnology development.

In the United States, more specific regulatory standards emerged in 1984 after the White House Office of Science and Technology Policy initiated work on what would become the 1986 Coordinated Framework for Regulation of Biotechnology. The latter established that biotechnology products would be assessed, not on the basis of their production methods (i.e. whether they were the result of genetic manipulation), but on the nature of their product traits. Therefore, new GM plants did not require any exceptional regulation if their traits did not differ from those of existing ones. Secondly, it granted regulatory oversight to the federal government in matters of biotechnology. In a federal system, this was a key step, removing potential conflict between the central government and the states, which could fragment the regulatory horizon.

As a result, in the United States, the promotional path to biotechnology development began early on. According to Krimsky (2005), the Reagan administration began to immediately frame risk assessment of biotechnology in a reductionist fashion, delivering regulatory oversight to the federal level and choosing scientists to define the contents of that regulation. The capacity of certain groups to participate in the debate about biotechnology (especially against it) was carefully shaped in the US, thus framing the issue so as to foster a more permissive approach (Jasanoff 2005). Furthermore, the scientific and commercial sides of the biotechnology sector and its national development were closely linked with a strong international push for the approval of favourable trade and intellectual property rules, which would ensure that American biotechnology products could find their way to foreign markets.

This vision influenced at least some of the decisions that were taken at the international level, helping to shape the approaches adopted in other countries. The publication by the OECD of its blue book on 'Recombinant DNA Safety Considerations' (OECD 1986) was a critical step in the diffusion of these standards to other countries. It was followed in the next ten years by a variety of similar publications in cognate areas (Schiemann 2006). Ad hoc organisations like the United States National Institute of Health Recombinant DNA Advisory Committee helped develop these standards, publishing the very first set of guidelines for the safe handling of recombinant DNA. The emergence of the substantial equivalency risk assessment principle (Canada 2001: 11), and the adoption of a product-based approach to regulation, were crucial steps in the evolution of these regulatory regimes (Murphy and Levidow 2006) and are now used in many countries to designate GM and other genomics products as equivalent to other non-GMO products that come to market outside of the biotechnology stream and thus avoid further scrutiny or regulation.

The initial phase of the biotechnology cycle in Europe also centred on the concerns expressed in the Berg Letter and at the Asilomar conference. Most European countries and the EU emulated the substantial equivalency model set out in previous FDA rulings. A discussion phase regarding the creation of the European Molecular Biology Organization (EMBO) debated the pre-regulatory and the standard setting phases of biotechnology, especially as it related to GMOs (Gottweiss 2005). As part of its increased activity in environmental and biotechnology regulation, the EEC produced *A Community Framework for the Regulation of Biotechnology* [COM (1986) 573] in November 1986. This aimed at creating comparable national standards in the sector for all members of the EEC, and looked at specific standards for the release of GMOs in the environment, assuming that this release was potentially dangerous. European institutions would spend the next four years debating the appropriate structure of this regulatory model.

Ultimately the strength of the European bureaucracy and the increased political salience of genetic modification crystallised the precautionary approach into the more direct regulation of the 1990 Directives 90/219 (Contained Use) and 90/220 (Deliberate Release), which set strict limitations upon GMOs (Gottweiss 2005).

Beginning with the controversy over transgenic maize of the mid-1990s, European publics increasingly delegitimised the introduction of transgenic material (Murphy and Levidow 2006), solidifying the divergence in the approaches followed in the US and Europe, with the EU moving beyond soft regulation to the beginnings of a more mature direct regulation system.

In mid-1998, EU Directive 98/44 on the Legal Protection of Biotechnology Inventions harmonised national legislation, most notably allowing the patenting of plants and animals. However, Article 95 of the Treaty of Amsterdam of 1997 institutionalised the precautionary principle and allowed member states to legislate limits to biotechnology applications and dissemination for the protection of the health of their citizens and of the environment even if this ran counter to free trade. A 1999 court challenge launched by Holland, Italy, and Norway in front of the European Court of Justice led to the decision of 24 June 1999 by the EU Council of Ministers to implement a de facto moratorium on GMOs. In 2000, the Final Report of the Biotechnology Consultative Forum argued that some kind of labelling for GM products was needed and also agreed on suggesting that limits be imposed on intellectual property rights.

This meant that the US and the EU were locked in a trade conflict and increasingly a transatlantic coalition of business and political actors began to develop to foster the acceptance of biotechnology in the EU, while NGOs advocated against it (Murphy and Levidow 2006). Nevertheless, the EU continued to develop direct regulation, passing Directive 2001/18 on the deliberate release of GMOs, which, notwithstanding very tight regulation, became the first concrete step towards a lifting of the 1999 moratorium. At the same time, it created the European Food Safety Authority (EFSA), which was tasked with safeguarding the health of consumers in the member states, and to undertake risk assessment and management of food and feed chains.

In 2004, the European Union officially lifted its 1999 ban, mandating member states to pass coexistence laws harmonising domestic and EU legislation. However, member states did very little to harmonise internal regulations to the EU position and still effectively stopped the introduction of GMOs. This position was condemned by the WTO (World Trade Organization) in 2006 as a breach to free trade, but very little has changed as enormous disincentives exist in the EU regulatory and political systems against the introduction of GM products. To begin with, public opinion, at least since the 1990s, tends to be very much against these products (Tiberghien 2009), especially as far as food is concerned (Gaskell et al. 2004). Second, Regulation 1830/2003/ EC regarding the labelling of GM products is among the strictest in the world. The regulation established a 0.9 per cent ceiling for GM content above which products must be labelled as containing GMOs. However, under pressure from public opinion, some member states adopted even stricter rules and moved to close a regulatory loophole that left eggs, dairy, and meat derived from animals that had been fed genetically modified animal feed out of the labelling scheme. These differences among member states reflect, and are at least partially explained by, the interstate competition inherent in EU governance (Tiberghien 2009). Finally, discrepancies in the reception of European decisions by member states further reduce the viability of these products in a system that is already very differentiated (Varone 2007). Table 4.8 contrasts the evolution of GM biotechnology regulation in the US and EU using a genealogical model.

Table 4.8 The GM biotechnology lifecycle in the US and EU

Bernstein Life-cycle Stage	Phase	USA	EEC/EU
I. Birth	Pre-regulatory	Berg Letter (1974); Establishment of the NIH Recombinant DNA Advisory Committee (Oct. 1974); Asilomar Conference (1975).	Berg Letter (1974); Asilomar Conference (1975); Discussion surrounding the European Molecular Biology Organization (EMBO) regarding recombinant DNA.
II. Infancy	Co-optive	NIH Guidelines for Research Involving Recombinant DNA Molecules (1976) and successive amendments especially 1978; Transfer to the EPA of the responsibility for assessing field released risks.	Attempt at adapting the US NIH Recombinant DNA Guidelines to the European context.
III. Early Childhood	Early Scientific Phase (1–2 years)	US Supreme Court on the patenting of GMOs [Diamond v. Chakrabarty, 447 U.S. 303 (1980)]; Work of the White House Office of Science and Technology Policy (1984–85); White House Coordinated Framework for Regulation of Biotechnology (1986).	Creation of the Biotechnology Steering Committee (1984); European Commission *A Community Framework for the Regulation of Biotechnology* (1986); EC Directive 87/22/EEC on the market approval of biotechnologies (Dec. 1986).
IV. Late Childhood	Middle Technical Phase (8–10 years)	Report of the National Academy of Science (1987); FDA Statement of Policy: Foods Derived from New Plant Varieties (May 1992); FDA Guidance for Industry: Voluntary Labelling for GM foods (2001)	Council Directive 90/220/EEC on the deliberate release into the environment of genetically modified organisms (1990); EC Directive 93/41/EEC on the market approval of biotechnologies (1993); Article 95 of the Treaty of Amsterdam – institutionalises the precautionary principle (1997); Directive 98/44/EC on the Legal Protection of Biotechnology Inventions – allows the patenting of GMOs (1998).

(Continued)

Table 4.8 (Continued)

Bernstein Life-cycle Stage	Phase	USA	EEC/EU
V. Adolescence	Late Administrative Phase (1–2 years)		Directive 2001/18/EC on the deliberate release into the environment of genetically modified organisms (2001); Creation of European Food Safety Authority (2002); Regulation 1830/2003/EC on the traceability and labelling of genetically modified organisms and of food containing GMOs (2003).

Conclusion

Other authors have attempted to explore regional differences in regulation (Isaac 2002) and understand the consequences of the polarised nature of biotechnology policy in different areas like North America and Europe (Bernauer 2003). Many variables have been invoked to explain the nature of regulatory differences, especially between the US and the EU, from the nature of risk perception (Isaac 2002; Toke 2004), to institutional and participation elements (Jasanoff 2005), to an issue of global governance (Murphy and Levidow 2006). As Kinchy *et al.* (2008) demonstrated, simple explanations, like that of a global pressure towards similar regulation, fail. An approach rooted in a more detailed study of regulation and regulatory regime lifecycles is required to sort out both the pattern of regulation and whether it reflects convergent or divergent pressures.

In conclusion, we can say that the complexities and fluidity of the field of biotechnology are captured by the twin components (spatial and temporal) of the comparative genealogical framework established here. As the model shows, significant differences exist between countries, both in terms of the substantive and procedural nature of the regimes they have adopted, as well as their temporal evolution. This two-tiered approach allows us to include a variety of both endogenous and exogenous variables like the pressures from transnational networks of the type highlighted by Murphy and Levidow (2006), and the endogenous variables described by Isaac (2002) and Jasanoff (2005) in their work on the subject. Furthermore, bringing together the spatial and temporal dimensions of regulation allows us to include a dynamic analysis of the phenomena.

Appendix 1

Table 4.9 The regulatory issue field in biotechnology

Biotechnology Activities	Regulatory Concerns				
	Research-Related	Legal	Economic	Education-Related	Acceptance- and Implementation-Related
Intellectual property rights	Patent policy	Intellectual property and licensing practices	Cost-effectiveness		Acceptance of biotech private ownership
Public information and Deliberative activities	Ethics review	Privacy and confidentiality	Cost of broad consultations	Development of clinical guidelines	Behaviour modification in response to biotechnology results
			Intellectual property	Classroom education	
				Public education	
				Risk communication	

(Continued)

Table 4.9 (Continued)

Biotechnology Activities	Regulatory Concerns				
	Research-Related	Legal	Economic	Education-Related	Acceptance- and Implementation-Related
Retail and Trade Activities	Patent law	Trade agreements	Market value and pricing Supply and demand Commercialisation of public-sector initiatives Creation of new market segments	Labelling	Public adoption of biotechnology
Health and Safety	Creation of a regulatory framework	Regulatory oversight (product and manufacturing review, labelling, laboratory quality and environmental impact)	Costs related to testing	Education of health professionals	Acceptance of the safety of food products by the public
Consumer Activities	Media advertising	Genetic discrimination	Different responses in consumer behaviour	Information directed towards consumers	Cultural respect

(Continued)

Table 4.9 (Continued)

Biotechnology Activities	Regulatory Concerns				
	Research-Related	Legal	Economic	Education-Related	Acceptance- and Implementation-Related
Research Investment Activities	Prioritisation of research areas (basic, applied and technology development) Allocation of funds Provision of facilities Access to tools and research samples	Protection of human subjects Ownership of research results	Research and development funding Economic incentives for biotechnology research	Information directed towards citizens	Acceptance of the value of biotechnology investment
Commercialisation Activities	Reliance on private or public generated research- Patent policy	Intellectual property rights	Accessing venture capital Creation of technology licensing organisations	Labelling Pedagogical research	Acceptance of the value and safety of biotechnology products Public opinion research

Source: Haga and Willard (2006: 967)

Notes

1 Types of voluntary regulation include legislated compliance plans; regulatory exemption programmes; government–industry negotiated agreements; certification; challenge programmes; design partnerships; and standards auditing and accounting (Moffet and Bregha 1999). While non-governmental entities may, in effect, regulate themselves, they typically do so, as Gibson notes, with the implicit or explicit permission of governments (Ronit 2001; Gibson 1999). As long as these private standards are not replaced by government enforced ones, they represent the acquiescence of a government to the private rules, a form of delegated regulation (Haufler 2000 and 2001; Knill 2001; Héritier and Lehmkuhl 2008; Héritier and Eckart 2008). As a 'public' policy instrument, self-regulation requires some level of state action, either in supporting or encouraging development of private self-regulation or retaining the 'iron fist' or the threat of 'real' regulation if private behaviour does not change (Cashore 2002; Gibson 1999; Porter and Ronit 2006; Cutler *et al.* 1999).

2 On a spatial level, contemporary agri-food related biotechnology regulatory regimes are closely connected with global or transnational frameworks like the Cartagena Protocol on Biosafety (Newell 2008), or the Codex Alimentarius for food safety (Lindner 2008), or multi-level regulatory ones such as those developed in the EU relating to GMOs (Falkner 2007). International regulation of biotechnology is relatively complex, covering the areas of arms control, health and disease control, environmental protection, trade, drugs controls and social impacts (Rhodes 2009:178–9). It also passes through a variety of international organisations with direct regulatory control over the field including the Codex Alimentarius Commission (CAC), the Food and Agriculture Organization (FAO), the Convention on Biodiversity Secretariat (CBDS), the United Nations General Assembly, the WTO, the World Health Organization (WHO) and the World Intellectual Property Organization (WIPO). Overall, almost forty regulatory instruments are in place to deal with the many facets of this regulation (Rhodes 2009). While these international and transnational factors provide some common features, regulatory regimes for specific biotechnologies are constructed around different technological and functional contexts and are influenced by additional idiosyncratic factors such as economic viability, public opinion, and consumer responses to new technologies and processes within specific national contexts (Montpetit 2005; Falkner 2007). As a result, despite some similarities, different countries regulate, foster, and support specific biotechnologies in different ways (Lindner 2008).

Bibliography

Aizenman, Joshua (2009) 'Financial Crisis and the Paradox of Under- and Over-Regulation', SSRN eLibrary.

Baldwin, Robert and Martin Cave (1999) *Understanding Regulation: Theory, Strategy and Practice*. Oxford: Oxford University Press.

Berg, Paul, David Baltimore, Herbert W. Boyer, Stanley N. Cohen, Ronald W. Davis, David S. Hogness, Daniel Nathans, Richard Roblin, James D. Watson, Sherman Weissman and Norton D. Zinder (1974) 'Potential Biohazards of Recombinant DNA Molecules', *Science* 185: 303.

Bernauer, T. (2003) *Genes, Trade and Regulation: The Seeds of Conflict in Food Biotechnology*. Princeton: Princeton University Press.

Bernstein, Marver H. (1955) *Regulating Business by Independent Commission*. Princeton: Princeton University Press.

Borraz, Oliver (2007) 'Risk as Public Problems', *Journal of Risk Research* 10(7): 941–57.

Boucher, Patrick M. (2008) *Nanotechnology: Legal Aspects*. Boca Raton: CRC Press.

Bowman, Diana M. and Graeme A. Hodge (2007) 'Nanotechnology and Public Interest Dialogue: Some International Observations', *Bulletin of Science, Technology & Society*. 27(2): 118–32.

Brockman, Joan (1998) ''Fortunate Enough to Obtain and Keep the Title of Profession: Self-Regulating Organizations and the Enforcement of Professional Monopolies', *Canadian Public Administration*, 41(4): 587–621.

Canada (2001) *Action Plan of the Government of Canada in Response to the Royal Society of Canada Expert Panel Report*. Ottawa: Queen's Printer.

Cantley, Mark (2007) *An Overview of Regulatory Tools and Frameworks for Modern Biotechnology: A Focus on Agro-Food*. Paris: OECD.

Cashore, Benjamin (2002) 'Legitimacy and the Privatization of Environmental Governance: How Non State Market-Driven (NSMD) Governance Systems Gain Rule Making Authority', *Governance*, 15(4): 503–29.

Cashore, B., G. Auld and D. Newsom (2003) 'Forest Certification (Eco-Labeling) Programs and their Policy-Making Authority: Explaining Divergence among North American and European Case Studies', *Forest Policy and Economics*, 5(3): 225–47.

Cismaru, Magdalena and Anne M. Lavack (2007) 'Canadian Tobacco Warning Labels and the Protection Motivation Model: Implications for Canadian Tobacco Control Policy', *Canadian Public Policy/ Analyse de Politiques*, 233(3): 1–11.

Crew, Michael and David Parker (2006) *International Handbook on Economic Regulation*. Cheltenham: Edward Elgar.

Cutler, Claire A., Virginia Haufler and Tony Porter (eds) (1999) *Private Authority and International Affairs*. Albany: State University of New York Press.

de Vries, Michiel S. (2010) *The Importance of Neglect in Policy-Making*. Basingstoke: Palgrave Macmillan.

Elgie, Robert (2006) 'Why do Governments Delegate Authority to Quasi-Autonomous Agencies? The Case of Independent Administrative Authorities in France', *Governance*, 19(2): 207–27.

Elliott, C. and R. Schlaepfer (2001) 'The Advocacy Coalition Framework: Application to the Policy Process for the Development of Forest Certification in Sweden', *Journal of European Public Policy*, 8(4): 642–61.

European Commission (2002) *Life Sciences and Biotechnology. A Strategy for Europe*. Brussels: European Communities.

Falkner, Robert (2007) 'The Political Economy of 'Normative Power' Europe: EU Environmental Leadership in International Biotechnology Regulation' *Journal of European Public Policy* 14(4): 507–26.

Fleischer, J. (2010) 'Europeanisation by agencification: How isomorphic EU agencies accomplish the European regulatory state', paper presented at the *EGPA Annual Conference*, Study Group XIV: EU administration and multi-level governance, 08–10 September 2010, Toulouse.

Furger, F., M. S. Garfinkel, D. Endy, G. L. Epstein and R. M. Friedman (2007) 'From Genetically Modified Organisms To Synthetic Biology: Legislation in the European Union, in Six Member Countries, and in Switzerland', in *Working Papers for Synthetic Genomics: Risks and Benefits for Science and Society*: 165–84.

Gaskell, George, N. Allum, W. Wagner, N. Kronberger, H. Torgensen and J. Bardes (2004) 'GM foods and the misperception of risk perception', *Risk Analysis* 24, no. 1: 183–92.

Gibson, Robert B. (1999) *Voluntary Initiatives and the new Politics of Corporate Greening*. Peterborough: Broadview Press.

Gilardi, Fabrizio (2005) 'The Institutional Foundations of Regulatory Capitalism: The Diffusion of Independent Regulatory Agencies in Western Europe', *Annals of the American Academy of Political and Social Science*, 598: 84–101.

Gottweiss, Herbert (2005) 'Transnationalizing Recombinant-DNA Regulation: Between Asilomar, EMBO, the OECD, and the European Community', *Science as Culture* 14(4): 325–38.

Haga, S. B. and Willard, H. F. (2006) 'Defining the Spectrum of Genome Policy', *Nature Reviews Genetics*, 7: 966–72.

Haufler, Virginia (2000) 'Private Sector International Regimes', in Richard A. Higgott, Geoffrey R. D. Underhill and Andreas Bieler (eds) *Non-state Actors and Authority in the Global System*. London: Routledge (121–137).

Haufler, Virginia (2001) *A Public Role for the Private Sector: Industry Self-Regulation in a Global Economy*. Berkeley: Carnegie Endowment for International Peace.

Héritier, Adrienne and Sandra Eckert (2008) 'New Modes of Governance in the Shadow of Hierarchy: Self-regulation by Industry in Europe', *Journal of Public Policy* 28(1): 113–38.

Héritier, Adrienne and Drik Lehmkuhl (2008) 'Introduction: The Shadow of Hierarchy and New Modes of Governance', *Journal of Public Policy* 28, no. 1: 1–17.

Hodge, Graeme A., Diana Bowman and Karinne Ludlow (2007) 'Introduction: Big Questions for Small Technologies', in Graeme Hodge, Diana Bowman and Karinne Ludlow (eds), *New Global Frontiers in Regulation: The Age of Nanotechnology*, Cheltenham: Edward Elgar (3–28).

Holzinger, Katharina, C. Knill and T. Sommerer (2008) 'Environmental Policy Convergence: The Impact of International Harmonization, Transnational Communication, and Regulatory Competition', *International Organization* 62: 553–87.

Hood, Christopher (1986) *The Tools of Government*. Chatham: Chatham House Publishers.

Howlett, M. and A. Migone (2010) 'Explaining Local Variation in Agri-food Biotechnology Policies: "Green" Genomics Regulation in Comparative Perspective', *Science and Public Policy*, 37(10): 781–95.

Isaac, Grant E. (2002) *Agricultural Biotechnology and Transatlantic Trade: Regulatory Barriers to GM Crops*. Oxford: Oxford University Press.

Jacobzone, Stephane (2005) 'Independent Regulatory Authorities in OECD Countries: An Overview', in OECD Working Party on Regulatory Management and Reform (ed.) *Designing Independent and Accountable Regulatory Authorities for High Quality Regulation – Proceedings of an Expert Meeting in London, United Kingdom 10–11 January 2005*. Paris: OECD (72–100).

Jasanoff, Sheila (2005) *Designs on Nature: Science and Democracy in Europe and the United States*. Princeton: Princeton University Press.

Kinchy, Abby J., Daniel Lee Kleinman and Robyn Autry (2008) 'Against Free Markets, Against Science? Regulating the Socio-Economic Effects of Biotechnology', *Rural Sociology* 73(2): 147–79.

Knill, Christoff (1998) 'European Policies: The Impact of National Administrative Traditions', *Journal of Public Policy*, 18(1): 1–28.

Knill, Christoff (2001) 'Private Governance Across Multiple Arenas: European Interest Associations as Interface Actors', *Journal of European Public Policy* 8(2): 227–46.

Krimsky, Sheldon (2005) 'From Asilomar to Industrial Biotechnology: Risks, Reductionism and Regulation', *Science as Culture* 14 (4): 309–23.

Kuhlmann, Ellen and Judith Allsop (2008) 'Professional Self-Regulation in a Changing Architecture of Governance: Comparing Health Policy in the UK and Germany', *Policy & Politics* 36(2): 173–89.

Kuzma, Jennifer and Todd Tanji (2010) 'Unpacking Synthetic Biology for Oversight Policy', *Regulation & Governance* 4(1): 92–112.

Leiss, William (2001) *In the Chamber of Risks: Understanding Risk Controversies.* Montreal: McGill-Queen's University Press.

Lindner, L. F. (2008) 'Regulating Food Safety: The Power of Alignment and Drive towards Convergence', *Innovation: The European Journal of Social Science Research* 21:133–43.

Majone, Giandomenico (2005) *Dilemmas of European Integration.* Oxford: Oxford University Press.

Majone, Giandomenico (2010) 'Foundations of Risk Regulation: Science, Decision-Making, Policy Learning and Institutional Reform', *European Journal of Risk Regulation* 1(1): 5–19.

Mashaw, Jerry L. (1988) 'Review: Mendeloff's The Dilemma of Toxic Substance Regulation: How Overregulation Causes Underregulation', *The RAND Journal of Economics* 19(3): 489–94.

May, Peter J. (2002) 'Social Regulation' in Lester Salamon (ed.) *Tools of Government: A Guide to the New Governance.* Oxford: Oxford University Press (156–85).

Moffet, J. and F. Bregha (1999) 'Non-Regulatory Environmental Measures', in R. B. Gibson (ed.) *Voluntary Initiatives: The New Politics of Corporate Greening.* Peterborough: Broadview Press (15–31).

Montpetit, Éric (2005) 'A Policy Network Explanation of Biotechnology Policy Differences between the United States and Canada', *Journal of Public Policy* 25 (3): 339–66.

Murphy, Joseph and Les Levidow (2006) *Governing the Transatlantic Conflict over Agricultural Biotechnology. Contending Coalitions, Trade Liberalisation and Standard Setting.* London: Routledge.

Newell, P. (2008) 'Lost in Translation? Domesticating Global Policy on Genetically Modified Organisms: Comparing India and China', *Global Society* 22: 115–36.

Nichols, Albert L. and Richard J. Zeckhauser (1988) 'The Perils of Prudence: How Conservative Risk Assessments Distort Regulation', *Regulatory Toxicology and Pharmacology* 8(1): 61–75.

OECD (1986) *Recombinant-DNA Safety Considerations.* Paris: OECD.

OECD (2009) *The Bioeconomy to 2030: Designing a Policy Agenda.* Paris: OECD.

Otway, Harry and Jerome Ravetz (1984) 'On the Regulation of Technology: Examining the Linear Model', *Futures* 16(3): 217–32.

Paarlberg, R. L. (2000) *Governing the GM Crop Revolution. Policy Choices for Developing Countries.* Washington, DC: International Food Policy Research Institute.

Porter, Tony and Karsten Ronit (2006) 'Self-Regulation as Policy Process: The Multiple and Criss-Crossing Stages of Private Rule-Making', *Policy Sciences* 39(1): 41–72.

Reagan, Michael D. (1987) *Regulation: The Politics of Policy.* Boston: Little, Brown and Company.

Rhodes, Catherine (2009) 'Is the International Regulation of Biotechnology Coherent?', *Journal of International Biotechnology Law* 6(5): 177–91.

Ronit, Karsten (2001) 'Institutions of Private Authority in Global Governance: Linking Territorial Forms of Self-Regulation', *Administration and Society* 33: 555–78.

Rosenbloom, David H. (2007) 'Administrative Law and Regulation', in Jack Rabin, W. Bartley Hildreth, and Gerald J. Miller (eds) *Handbook of Public Administration*. London: CRC Taylor & Francis (635–96).

Schlager, Edella (1999) 'A Comparison of Frameworks, Theories, and Models of Policy Processes', in Paul A. Sabatier (ed.), *Theories of the Policy Process* (2nd edn). Boulder: Westview Press.

Schiemann, Joachim (2006) 'The OECD Blue Book on Recombinant DNA Safety Considerations: Its Influence on ISBR and EFSA Activities', *Environmental Biosafety Research* 5(4): 233–5.

Scholz, John T. (1991) 'Cooperative Regulatory Enforcement and the Politics of Administrative Effectiveness', *American Political Science Review* 85(1): 115–36.

Seaton, Anthony, Lang Tran, Robert Aitken and Kenneth Donaldson (2009) 'Nanoparticles, Human Health Hazard and Regulation', *Journal of The Royal Society Interface*. 7(1): 1–11.

Sheingate, A. D. (2006) 'Promotion versus Precaution: The Evolution of Biotechnology Policy in the United States', *British Journal of Political Science* 36(2): 243–68.

Sinclair, D. (1997) 'Self-Regulation Versus Command and Control? Beyond False Dichotomies', *Law and Policy* 19, no. 4: 529–59.

Stern, Jon (1997) 'What makes an independent regulator independent?', *Business Strategy Review* 8(2): 67–75.

Tiberghien, Yves (2009) 'Competitive Governance and the Quest for Legitimacy in the EU: The Battle over the Regulation of GMOs since the Mid-1990s', *Journal of European Integration* 31(3): 389–407.

Toke, David (2004) *The Politics of GM Food: A Comparative Study of the UK, USA and EU*. London: Routledge.

Trebilcock, Michael J. (2008) 'Regulating the Market for Legal Services', *Alberta Law Review* 45 (2008): 215–32.

Trebilcock, Michael J., C. J. Tuohy and A. D. Wolfson (1979) *Professional Regulation: A Staff Study of Accountancy, Architecture, Engineering, and Law in Ontario*. Prepared for the Professional Organizations Committee. Toronto: Ministry of the Attorney General.

Tuohy, Carolyn and Alan Wolfson (1978) 'Self-Regulation: Who Qualifies?', in Slayton, Philip and Trebilcock Michael J. (eds), *The Professions and Public Policy*. Toronto: University of Toronto Press.

Varone, Frédéric (2007) 'Comparing Biotechnology Policy in Europe and North America: A Theoretical Framework', in É. Montpetit, C. Rothmayr Allison and F. Varone (eds) *The Politics of Biotechnology in North America and Europe: Policy Networks, Institutions and Internationalization*. Lanham: Lexington.

Vick, D. W. (2006) 'Regulatory Convergence?', *Legal Studies* 26(1): 26–64.

5   Pragmatism revisited

An overview of the development of regulatory regimes of GMOs in the European Union[*]

Anders Johansson

Introduction

The regulation of biotechnology has, from the early days of the 1970s, been preoccupied with the problem of how to establish legitimate and reliable regulations. Hence, political bewilderment is a key characteristic of the area of GMOs. The uncertainties of risks in biotechnological innovations have made scientific knowledge essential for decision makers in the area of GMOs. Political decision making is thus exposed to the very same scientific uncertainty that it tries to diminish in regulatory decisions concerning GMOs. The political discourse that has emerged as a consequence of establishing a European regulatory framework for GMOs has not been without problems. During the first struggling efforts in the early 1980s, the GMO legislation has experienced serious crises. As a result, it has been rearticulated, reshaped, and reformulated in conjunction with a changing discursive 'landscape' where issues such as, for example, risks, precaution, labelling, and coexistence have been dominant concepts in the discourse.

The use of potentially dangerous technologies in contemporary societies has increased the regulatory pressure on these technologies and brought about new demands on decision-making institutions. Controversies in the regulatory process regarding contested technologies emerge from two different discourses. The first concerns scientific uncertainties in the predictions of hazardous consequences and the second is the inherently normative character of the technologies. Previous studies have indicated that since the European Commission's regulatory effort regarding biotechnological innovations in the early 1980s, ethical, social, and democratic issues have become increasingly entangled with policy making for GMOs. The unfolding of European GMO politics has been a lesson in how policymakers try to reach equilibrium between the scientific and normative dimensions of these technologies in order to create a regulatory framework capable of obtaining the necessary social and political legitimacy. The development of regulatory regimes for GMOs in the EU is an example of how contemporary conceptions of biotechnology are an outcome of social, political, economic, and scientific interpretations and strategies deployed by various actors involved in the regulation of that specific technology.

In European GMO politics, policymakers have struggled with the notion of 'risk'. Brian Wynne has argued that the precedence of framing risk issues in scientific terms has two major effects (Wynne 1996). The first is that the experts do not incorporate any socially valued dimensions in the risk concepts. Examples of socially valued risk dimensions are whether the risks are voluntary or not, or if they are reversible or not. These issues have a strong impact on how the public regards risks. The other aspect concerns how experts and the public make different assumptions about the actual risk situation. For example, experts may take for granted the trustworthiness and competence of controlling bodies, whereas the public tends to question the institutions concerned.

In this chapter I will reconstruct the regulatory trajectory of GMOs at the EU level during four episodes in time. Each episode led to distinct regulatory structures, which can be understood as different 'regulatory regimes'. The four episodes are:

1   The creation of the Deliberate Release Directive 90/220/EEC (1985–90).
2   Regulatory development during the European GMO moratorium (1996–2004).
3   The creation of Directive 2001/18EC and the ensuing regulatory process (1997–2001).
4   The debate about the coexistence and segregation of GM crops (2001–07).

I will then relate the above regulatory episodes to different policy models to provide an account of the regulatory transformations that have occurred. For an overview of key events in GMO regulation in the EU see Table 5.1. The events that I focus on here are emphasised. Note specifically the overlapping timeframes of the European GMO moratorium (1996–2004) with the regulatory process of Directive 2001/18 (1997–2001).

Table 5.1 Key events in EU biotech regulation

Date	Events
1983	Biotechnology is included in the EU Framework R&D Programme
1984	The EU Commission forms the Biotech Steering Committee
1986	EU Commission Report: *A community Framework for the Regulation of Biotechnology*
1990	**EU Council adopts Directives 90/219 and 90/220**
1996 (March)	Start of the BSE crisis leads to questioning of EU food safety regulation
1997 (Jan)	EU Council adopts Novel Food Regulation
1997 (Jan)	EU Commission approves the sale of GM maize; three member states invoke the safeguard clause
1998 (Oct)	Start of the *de facto* moratorium on approval of new GM varieties

(Continued)

Table 5.1 (Continued)

Date	Events
1999 (June)	**Declaration of moratorium on GM approvals by five member states**
2000 (Jan)	Commission adopts White paper on Food Safety
2000 (Jan)	Commission adopts White paper on the Precautionary Principle
2000 (Jan)	The international agreement Cartagena Protocol on Biosafety is adopted and the Protocol entered into force on 11 September 2003.
2001 (March)	**EU Council and European Parliament [EP] adopt Directive 2001/18, replacing 90/220, on deliberate release of GMOs**
2002 (Jan)	Establishment of the European Food Safety Authority
2002 (Jan)	EU Commission adopts a Communication entitled *Life Sciences and Biotechnology: A Strategy for Europe.*
2003 (May)	US launches WTO complaint over EU regulation of GMOs
2003 (July)	**EU Commission publishes *Recommendation on Guidelines For Coexistence***
2003 (Sept)	EU Council and EP adopt Regulation 1829/2003 on Genetically Modified Food and Feed
2003 (Sept)	EU Council and EP adopt Regulation 1830/2003 on Traceability and Labelling
2004 (Apr)	Entry into force of Regulations 1829/2003 and 1830/2003
2004 (May)	EU Commission ends moratorium with approval of Bt-11 maize
2006 (May)	WTO publish their decision that the EU's moratorium was illegal
2007 (April)	EU Commission publishes its review of life sciences and biotech strategy. The review proposes to refocus EU's strategy plan on five interdependent priority actions

Regulatory regimes in European GMO politics

The rDNA controversy

The points of departure are the Asilomar conferences in United States during the early 1970s, known as the 'recombinant DNA controversy'. Those conferences did not address GMOs but instead a scientific predecessor known as recombinant DNA (rDNA) technology. It can be said that the Asilomar conferences became the regulatory heritage for genetic engineering both in the EU as well as among the member states. At the Asilomar conference in 1973, invited scientists met to review scientific progress in research on recombinant DNA molecules and techniques but also to discuss how to deal with potential biohazards of the novel research (Wright 1994). No consensus on any recommendations could be agreed on at the conference and later a moratorium for further research regarding some experiments in genetics was enforced.

Another Asilomar conference in 1975 differed from the first one in 1973 by the presence of media and lawyers. The consequence of inviting reporters was that the potential hazards and the scientific uncertainties of rDNA became public knowledge and a topic of debate in the media. Previously, the possible hazards of genetic engineering had been primarily an internal scientific concern about the potentially negative effects of incorporating new DNA in bacteria. The tension that was rising from new forms of knowledge in rDNA research within the scientific community regarding risks issues, revealed a horizon of uncertainties for both scientists and policymakers, which, in turn, opened up new possibilities for regulatory actions. The scientific community involved in rDNA research quickly realised that new regulatory actions could either hinder or promote further research depending on the outcomes.

Donald S. Fredrickson, the former Director of the National Institutes of Health (NIH), describes the scientist's reaction as a kind of reflex. He writes:

> Almost instinctively, the scientists at the fateful Gordon conference in 1973 had turned neither to the NIH nor to any other government agency, but first to the National Academy of Sciences (NAS), preferring to share their dilemma 'in the family,' seeking an 'in-house' solution (Fredrickson 2001).

It should, however, be noted that there were different opinions among the scientists at the 1975 conference. The majority of the Asilomar conference participants did agree that some kind of special guidelines were necessary and considered that there were valid reasons for concern because the technology could contain significant hazards (Watson and Tooze 1981). But there also existed a small but strong minority against any guidelines. Most prominent of these people were Stanley Cohen, Joshua Lederberg, and James Watson, who had expressed their opposition to the whole proceeding. Watson, for example, had 'proposed that the moratorium be lifted and no special precautions or guidelines be self-imposed' (ibid.). Hence, there existed a tension between those who argued for precaution and guidelines for experiments and those, like Watson, who were against any self-imposed guidelines.

Even if Watson and other leading scientists were against any guidelines in molecular biology they did not want to give the government any reason to interfere as it could interrupt further research in the field. Creating self-imposed guidelines became a sort of action to prevent such a possibility. The result from the conference was an agreement on how to proceed on two different groups of experiments – low risk and high risk. The moratorium was then lifted from these low risk experiments and, regarding the high risk experiments, it was argued that the risks could be controlled using available laboratory techniques. However, other scientists argued that the hazards involved in rDNA research had been underestimated and that the result from the conference more reflected personal interest among the scientists and the scientific community itself. The 'other' risk paradigm was represented by a small number of scientists who wished to restrict experiments with recombinant DNA in a much harder way and even wanted to prohibit recombinant DNA research but they became effectively marginalised.[1]

The process of making the Deliberate Release Directive 90/220/EEC

When the applications of genetic engineering shifted from laboratory research with recombinant DNA techniques to applied science with new biotechnological products intended for the market, the social and political dimensions of the technology had to be addressed by policymakers. Even if the political and social climate were significantly different when biotechnology was debated at the Asilomar conferences twenty years earlier, there were also some important resemblances. The resemblance can be found in the overall aim of establishing a regulatory and legislative framework in both recombinant DNA and for GMOs. In both the Asilomar case and when the EU started to create a regulatory framework for biotechnology the overall aim was to achieve control over a new and contested technology by arranging a framework where the risks could be controlled but still allowing the technology to be implemented and promoted.

When the governments in the EU member states started to develop their own regulatory framework it soon became clear that the regulatory styles differed profoundly within Europe. The divergence caused the EU to take action and try to harmonise the framework and, in 1990, the European Commission presented two new directives on biotechnology. It was the result of several years of long deliberations and negotiations. The first was the Directive on Contained Use, known as Directive 90/219 (Council of European Union 1990a), which is a regulation concerning the contained use of genetically modified microorganisms for both research purposes and for production. The contained use of microorganisms for research was the same topic that was discussed at the Asilomar conference and had now been regulated even in Europe. The other regulation was the Directive on Deliberate Release and Placing on the Market of Products, known as Directive 90/220 (Council of European Union 1990b).

The heritage from Asilomar was passed on through two different strands: the first was the regulatory heritage and was developed through the OECD report *Recombinant DNA safety considerations* from 1986 (OECD 1986). The report used the regulatory consequences from the Asilomar episodes even if it developed them further, and both European member states as well as the European Commission utilised the OECD report in order to create their own regulatory framework. The other influence was the cognitive framing, i.e. in Asilomar there was a lot of participation from European scientists, whom later were involved as regulatory experts in the European regulatory process in their own countries.

With the new directive, the EU Commission provided the first unifying legal framework for GMOs in Europe. It came to have trans-Atlantic consequences as well as consequences for the member states, which are obliged to integrate EU directives into their national legal frameworks. This was a new precautionary approach, that differed from the more industry-driven and scientific regulatory tradition that had been adopted in the USA. The policy framework implemented in the USA was based on the concept of 'substantial equivalence'. The contrast between the precautionary approach implemented by the EU vis-à-vis the concept of substantial equivalence can be summed up in the following way: the USA says yes to GMOs unless proven hazardous, whereas the EU says no unless they are proven safe. This difference is primarily ideological. The proceedings of the regulatory process leading to the European Directive 90/220 suggest that the protection of the market was more important, and the weight of evidence of harm to human health or the environment had to be obvious, present, and scientifically evidential (for extended discussion see Johansson 2009).

Even if the primary focus of Directive 90/220 was the harmonisation of the market, the directive represented a new legislative approach towards the regulation of GMOs. The directive included some elements that were considered to embrace a precautionary approach, in particular its focus on process-based evaluations of new biotechnologies. But it was mostly the new risk preventions strategies that caused most discussions. The European Commission introduced extensive risk assessment procedures and provided safety standards in an appendix to the directive. It was argued that the most important reason for implementing these risk prevention strategies was that the deliberate release of GMOs was still associated with uncertainties regarding the potential risks for humans and the environment.

In Directive 90/220, it was necessary for the European Commission to address the public concern regarding the risks in relation to GMOs. The Commission needed to achieve equilibrium between the public demand for a precautionary approach and the industrial demand for a purely science-based approach, and at the same time create a consistent and homogenous regulation that could be accepted across the EU. In Directive 90/220, the precautionary principle did not exist as a principle as such. Instead, the directive supported a precautionary *approach*. The precautionary approach can be seen as a weaker form of the precautionary principle and, therefore, it could be regarded as a more ideological then scientific standpoint. However, it is also a direct consequence of trying to incorporate both public concerns as well as demands from the biotech industry. Directive 90/220 was however acknowledged as a regulatory framework based on a precautionary approach because of the introduction of the new risk assessment procedures and the process-based approach, meaning that the legislation has targeted the GMO technology instead of a product-based approach that would focus on all products without consideration to how it was made. The regulation also introduced step-by-step and case-by-case evaluation procedures, meaning that before any market introduction of a GM crop it would have to be preceded by field trials (step-by-step) and that each crop would be assessed on an individual basis (case-by-case).

It could be argued that Directive 90/220 is a watered-down result since it is a compromise between meeting the demands of the market and of protecting human health and the environment. In order to reach some sort of equilibrium between these two objectives, the policymakers tried to be more sensitive towards the concerns expressed in the public sphere. The legislative approach was needed to respond to the regulatory challenges of: (1) making legitimate law and (2) adequately addressing the risks but also (3) had the prospect of being adopted by all member states independent of their positions on GMOs, which was the main concern for the European Commission. This was done by trying to adopt social, legal, political, and economic concerns within the regulation. The resulting regulatory regime in Europe was seen, on the one hand, by the market and biotech industry to be excessively inclined towards precaution, with demanding risk assessment procedures. On the other hand, it was seen by green NGOs and consumer groups as not precautious enough, with risk assessment procedures that were too simplified and reductionist.

The moratorium – towards a new cognitive framing of uncertainties

Following a European Council meeting in June 1999, France, Denmark, Greece, Italy, and Luxembourg declared that they would stop all new authorisations of GMOs. The member states argued for a more transparent regulatory framework and a more sophisticated risk assessment procedure to restore public and consumer confidence. Under pressure from these, as well as others member states, the European Commission initiated a moratorium on the production and sale of GM products when the decision for revising the criticised Directive 90/220 was taken. This marked a period of difficult policy issues and complex regulatory challenges for the EU. Member states would not accept the import of any products that could contain GMOs until a satisfactory legislation was developed. The first directive therefore lost its legitimate force to function as a regulatory framework for the deliberate release of GMOs.

During the European moratorium and the process of creating a revised directive on GMOs, there was an increasing need to build public confidence through more precautionary politics. The approach that resulted can be seen as a regulatory response to increasing uncertainty in GMO politics. Policy making in the EU became more open to a plurality of values expressed in the public sphere that had previously been excluded from the policy making process. Another result was that the second period constituted a further shift away from an industry-driven policy towards a much more consumer-sensitive policy.

The first indication of a more consumer-sensitive policy came in the Novel Food Regulation (European Parliament 1997). By stating that consumers should be informed about whether food products could contain substances that would fall into the category of 'ethical concerns', the regulation tried to resonate with normative concerns expressed in the public sphere. Furthermore, the legislation could be identified as a regulatory attempt to dissolve the tension and conflict that had started to arise in Europe regarding GMOs.

There are two main reasons the moratorium came about, both related to Directive 90/220. First, Directive 90/220 could be considered *ineffective* as it had failed to address such issues as labelling requirements or traceability of GMOs, which was important for the member states. Second, the directive did not succeed in obtaining the necessary *social legitimacy* because a number of relevant actors did not believe the directive was sufficiently anchored in the public sphere, or reflected the public sphere's normative concerns with GMO regulation. The hastening of the new Novel Food Regulation could not change this crisis, even when it addressed the issue of labelling. I would argue that the crisis in European

GMO politics was rooted both in the regulation's own ineffectiveness as well as in normative pressures that resulted from the regulation's failure to relate to the public sphere.

Directive 2001/18EC and the precautionary principle

When the revision of Directive 90/220 was announced in the *White Paper on Growth, Competitiveness, and Employment*, the Commission argued that 'such releases [of GMOs] have not given cause for concern, and evidence is accumulating to the effect that genetically modified plants do not differ from non-modified plants' (European Commission 1994). The Commission therefore wanted to introduce simplified procedures for promoting the GM industry and keep pace with the development in the USA. However, the regulatory process ultimately took a different turn. Towards the end of the regulatory process, the Commission made the following admission:

> In response to the growing public concerns about potential adverse effects of GMOs, the need for a more transparent and stringent regulatory system for the deliberate releases of genetically modified organisms into the environment has now become clear (European Commission 2000a).

The moratorium is an important factor for understanding the u-turn shift in the position of the Commission. That the Commission first argued for simplified procedures and at a later stage argued for the need of a more stringent regulatory system was a necessary concession in order to get the member states on track for a new regulatory framework on GMOs.

I would argue that the precautionary principle offered a way out of the predicament of balancing different interests from the GM industry and public concerns. The precautionary principle became a nexus where the regulations could interconnect with different actors' values and norms. The interpretative flexibility inherent in the precautionary principle should not be underestimated regarding its role in taking the EU out of the moratorium. For example, the interpretative flexibility made it possible for actors from different communities to use it as a common point of reference and then function as a means of coordination. The precautionary principle made this possible and brought into alignment the different social worlds inhabited by the GM industry, advocates of the internal market, NGOs, and consumers. But to achieve the interpretative plasticity in the precautionary principle, the regulatory process itself had to be opened up and made more participatory.

The objective of the Novel Food Regulation and Directive 2001/18 was to inspire public confidence in the regulatory framework for GMOs. The approach for achieving this was to make the regulatory framework increasingly more sensitive to the public sphere. It led to an increasing politicisation of the regulatory framework and a more consumer-sensitive legislation. Directive 2001/18 declared that: 'Member States, in accordance with the precautionary principle, shall ensure that all appropriate measures are taken to avoid adverse effects on human health and the environment which might arise from the deliberate release or the placing on the market of GMOs' (European Council 2001). The explicit reference to the precautionary principle is one of the major changes in the new directive. However, the explicit use of the principle became more difficult and revealed the problems with the inherent interpretative plasticity in the concept as the member states arrived at very different interpretations of the principle in their decision making. During the making of Directive 2001/18, the Commission published guidelines for interpreting and employing the precautionary principle in the EU (European Commission 2000b). That such a document came into existence suggests there was a need for the Commission to reduce the interpretative plasticity in the precautionary principle by reconstructing the concept such that a more uniform application of the principle resulted.

Policy making regarding the GMO coexistence controversy

The next conflict that emerged in the European regulation of GMOs concerned how segregation practices should be managed between GM and other crops, such as conventional and organic crops, in order to prevent admixtures. The coexistence controversy demonstrated that there still existed political differences between some member states and the Commission regarding GMOs. The controversy over coexistence measures also brought to the fore, again, the characteristics that had accompanied the regulatory landscape of GMOs – that is, high scientific uncertainty, lack of public trust, risk, diverse interests, and asymmetrical power among the actors involved.

The regulatory dispute between the European Commission and Upper Austria illustrates the regulatory frictions that emerged from the coexistence issue. I will use the example of the Commission vs. Upper Austria for several reasons. First, the conflict makes the different standpoints on the coexistence issues transparent. Second, Upper Austria tried to ban GMOs by justifying its decision based on EU law, which compelled the Commission to employ the legislation in an actual case. Third, Upper Austria also argued that they could present new evidence for the possibility of GMOs having negative consequences in their specific environments. Finally, through the study of Upper Austria it is possible to analyse differences and similarities in the Commission's response to the coexistence controversy, with respect to how the Commission responded to the moratorium episode. Taken together, the case is interesting for investigating different arguments about coexistence and the Commission's rationale.

Two different discourses on the controversy over coexistence and GMOs emerged. I refer to one of these as the 'coexistence discourse', which was promoted by the Commission; the other discourse is best characterised as the 'segregation discourse', which was supported by anti-GMO regions and local politicians in Europe, together with environmental NGOs. The case of the Commission vs. Upper Austria can be seen as an example of the confrontation of these two discourses, such that the arguments and standpoints had to be made explicit. The 'coexistence discourse' corresponded to the regulatory steering and governing rationale employed by the Commission to establish that no form of agricultural regime should be excluded in the European Union. The Commission advanced three arguments for applying different agriculture practices, namely freedom of choice for economic operators, the safeguarding of sustainability, and the diversity of agriculture in Europe (European Commission 2002: 21). However, disagreements about the safeguard measures for avoiding genetic mixture between the different agriculture regimes emerged immediately.

In the case of coexistence, three different issues provoked debate. First, on the issue of GMO-free zones, several member states declared that the Commission should define how GMO-free zones could be established. Second, tolerance thresholds for the adventitious presence of GMOs in seeds needed to be discussed in detail, especially the presence of GMOs in organic crops. Finally, there was the issue of liability in the event of contamination of conventional and organic crops by genetically modified crops.

To respond to these objections by member states and NGOs, the European Commission published guidelines for the development of strategies and best practices to ensure the coexistence of genetically modified crops with conventional and organic farming on 23 July 2003 (European Commission 2003). The recommendation's intention was to help member states develop methods for coexistence in agreement with EU legislation and thus overcome the confusion, ambiguity, and controversy that had surfaced over the implementation of coexistence rules. In the guidelines, the measures for ensuring the coexistence of the production of organic and conventional crops with GMOs were introduced, along with the declaration that member states would be allowed to take appropriate measures to ensure that there would be no unintended presence of GMOs in other products.

The declaration that the member states would be able to take action to prevent unwanted genetic flows was possibly an attempt to reassure the member states and avoid conflicts regarding coexistence issues. Nevertheless, opposition to the Commission policy emerged in some member states. Organised attempts to make GMO-free regions in EU utilising the 'segregation discourse' included arguments for the protection of farmers' autonomy and the right of member states to determine their own ways of farming, eating, producing, and selling food, as well as protecting the environment, seeds, and cultural heritage (GMO-free-Regions. org 2005). The movement to create GMO-free regions has been quite successful with over 230 regions and more than 4000 municipalities, including other local entities. Tens of thousands of farmers and food producers in Europe have declared themselves GMO-free, according to GENET.[2]

Returning to the European Commission vs. Upper Austria, the main controversy regarding Upper Austria's declaration to be GMO-free and ban the use of GMOs was about whether new scientific evidence existed for the likelihood of harmful consequences of using GMOs. Upper Austria argued that a new study proved that this was the case. The Commission requested that the European Food Safety Authority (EFSA) investigate whether the new information provided by the study would invalidate the environmental risk assessment in Directive 2001/18. The EFSA response was that there was no invalidation and that the study did not provide any new scientific evidence regarding risks to the environment and human health.

However, apart from EFSA's scientific opinion that Upper Austria did not provide any new evidence, the Commission also argued that the scientific findings in the study were published prior to the adoption of Directive 2001/18/EC. It is a legal necessity that findings are published *after* the adoption of Directive 2001/18 in order to be regarded as 'new'. Upper Austria, however, argued that it was the *scientific assessment* of the findings and data that was new. The European Commission also made other legal assessments in conflict with Upper Austrian demands. As a consequence, the Commission brought Austria to the European Court, which ultimately ruled that the Commission was right on all charges and that Austria had failed to fulfil its obligations under Directive 2001/18/EC.

As mentioned above, member states have continued to create GMO-free regions by invoking the safeguard clause, which enabled them to suspend the marketing or growth of GMOs that have EU-wide authorisation. The response by the European Commission was to disallow their demands and order them to lift their national bans. In 2008, the European Coexistence Bureau (ECoB) was created by the Directorate General for Agriculture and Rural Development (DG AGRI) and the Joint Research Centre (JRC). The ECoB is a direct result of the Council of Agricultural Ministers' adoption of the conclusion from a report on coexistence measures in May 2006 (Council of European Union 2006). The Council argued for the development of Community guidelines for crop-specific coexistence measures in cooperation with the member states. ECoB was then formed to help the member states by offering non-binding guidance on practices for technical segregation measures and to develop crop-specific guidelines for coexistence regulations. The new authority is responsible for the definition of coexistence measures, and is intended to help the member states develop their regulation on coexistence within these definitions.

On 2 April 2009, the Commission issued a second report on coexistence stating that the member states had made major improvement in developing coexistence legislation. The report also stated that the Commission did not see any indication for deviating from the subsidiarity-based approach on coexistence (European Commission 2009). However, already in March 2010, the European Commission announced that before the summer they would put forward a new proposal on how the EU science-based authorisation system could be integrated with the member states aspiration to decide autonomously on the cultivation of GM crops. On 13 July 2010, the Commission gave the full responsibility of GM cultivation to the member states (European Commission 2010b). The Commission also declared that Directive 2001/18 would be revised by a new paragraph that stated that member states would be able to restrict or prohibit GMO cultivation within their own territories without invoking the safeguard clause (European Commission 2010a). This new approach gave the member states the freedom to choose whether they want to allow, restrict, or ban the cultivation of GMOs within their territory.

Policy models in European GMO politics

The inverted decisionism model in Directive 90/220

According to the technical rationalism approach, risks are considered institutionally unproblematic for the policymakers, and the handling of risk can be exclusively dealt within an expert-dominated framework (Millstone 2007: 487–8). With respect to risk construction in the process of creating Directive 90/220, the norm-setting authorities were experts from the scientific community and interpreted who responded to risk issues by employing technical rationalism. The policy approach in the process of creating Directive 90/220 still employed the norms of technical rationalism but had in some sense moved beyond being solely a scientific enterprise. Although the norm-setting authorities were from the scientific community, Directive 90/220 also addressed economic, political, and normative issues: economic, because of the intention by policymakers to bring the GMO regulation within the market harmonisation framework; political, because the final Directive 90/220 was based on certain regulatory approaches to achieve the harmonisation goal; and normative, because it also included the precautionary approach.

In the standard model of decision making, policymakers first identify the goals of public policy and then scientific experts point out the best available means to those ends (Zwanenburg and Millstone 2005: 20). The inverted decisionism model of policy making is a model in which scientific experts define objectives and goals – for example regarding risk definitions, risk assessment procedures, and risk communication strategies – and policymakers subsequently decide on ways to fulfil the requirements of the science-based objectives. Millstone describes the inverted decisionism model in the following passage:

> Policy making could, however, still be represented as "science-based"; once scientists have decided what is, and would be, either safe or unsafe, responsibility would pass downstream to policymakers to decide how best to implement the advice of the scientists, taking account of non-scientific factors such as the cost of alternative courses of actions, and the ways different groups of protagonists might respond to policy signals and/or regulatory requirements (Millstone 2007: 493).

The above quote quite accurately describes how the policy approach with Directive 90/220 was conducted. However, as the period with the moratorium indicated, it would not be possible to uphold the dividing boundaries between science and policy in GMO policy making.

The Novel Food Regulation emerged from a regulatory 'anxiety' with the intention of being more 'consumer-friendly' while also taking the internal market into consideration. However, the reference in the regulation to 'ethical concerns' dissolved some of the regulatory 'tensions' about those concerns. The legislation was a regulatory move to dissolve the conflict that had begun to arise in Europe regarding GMOs. The inclusion of the idea that the consumer has a right to be informed about ethical concerns is an example of how public opinion formation in the 'weak' public sphere has been processed through the 'strong' public spheres of official political institutions. Hence, the Novel Food Regulation attempt to incorporate a more consumer-friendly approach could be understood as a move away from using the law solely as a steering medium for the enterprise of protecting the internal market, and instead increasing the validity of GMO regulations by relating it to normative components in the public sphere.

The moratorium could be interpreted as a process that changed the legislative approach from a techno–scientific framing towards an increased sensitivity to values of a moral and political nature, which, as I argued above, could be observed already in the Novel Food Regulation. The regulation was an effort to satisfy the need to restore public and market confidence regarding food safety and GMO regulation. Another contribution made with the Novel Food Regulation was that the legislation attempted to handle an increasing complexity and ambivalence in the GMO field with greater sensitivity towards the public sphere and by translating ethical concerns into the lawmaking realm. However, it also brings an increasing complexity and ambivalence into the regulatory process, as civil servants have to relate and interpret the meaning of 'ethical concerns'. Thus, the Novel Food Regulation may be a starting point for an increasing sensitivity towards the public sphere in Europe.

The co-evolutionary model in Directive 2001/18

The policy making model used in Directive 2001/18 may be a co-evolutionary model defined through reciprocal links between science and policy. According to Millstone:

> A key feature of this model is that it represents specific scientific delibera-
> tions as located in particular contexts, which have social, economic, ethical,
> and policy dimensions. The model is predicated on the assumption that those
> contexts affect the content and direction of those deliberations. Consequently
> representations of risks are hybrid judgments constructed out of both scien-
> tific and non-scientific considerations, even if they may be portrayed as if
> they were purely scientific (Millstone 2007: 498).

The policy approach involved in Directive 2001/18 could be seen as an exam-
ple of the above quote. It transformed the technical rationalist approach into a more constructive form of rationalism. In the co-evolutionary model, there are two important changes. First, it has more of a deliberative approach to policy making, which also was recognised in the process of making Directive 2001/18. Second, in the co-evolutionary model there is the concept of framing assumptions, which Millstone argues conceptualise the assumptions scientists make about objects and phenomena in which they are interested. Which objects and phenomena are relevant and which should be ignored? Furthermore, the concept

of framing assumptions also means that scientists make certain assumptions about what should count as relevant forms of knowledge and evidence, while supporting certain normative and interpretative patterns.

The European GMO moratorium created incentives to open up these framing assumptions for discussion, such as those concerning the precautionary principle and risks. The co-evolutionary model also indicates that at this stage of the regulatory process there is an increase in reciprocal communication, leading to difficulties regarding the regulatory steering of GMOs because of an increasing complexity and stress in the regulatory system from ongoing interpretation of this communication. In the case of the moratorium and the process of developing Directive 2001/18, it was the cognitive framing of uncertainties that would no longer fit within the Commission regulatory account. While public protests challenged the marginalising of accounts of uncertainties, most obviously regarding risk issues, there was an increasing need to reconstruct, for example, evidence of harm, comprehensions of the precautionary principle, and risk assessment procedures. It created a complexity in the regulatory system that resulted in increased communication in response to the increased complexity.

Millstone argues that the concept of 'framing assumptions' indicates that there is a set of broader considerations among policymakers and scientists that condition the production of scientific claims and results (Millstone 2007: 499). It is interesting to consider these conditions in the coexistence controversy. For example, Millstone argues that:

> In science-based risk debates, one of the main reasons why different groups of experts reach different conclusions is not because they reach conflicting interpretations of shared bodies of evidence but because they adopt differing assumptions about the categories of risk that they should address and those that they should discount or ignore. In other words, they reach different conclusions because they are answering different questions (Millstone 2007: 499).

If Millstone is right that the different 'framing assumptions' could lead to different conclusions about risks, it is tempting to use the same explanatory model in the coexistence controversy to address the separating distances between GM fields and other fields, the level of accepted GMO mixtures in other crops, and whether GMOs pose risks for human health or the environment.

The limited co-evolutionary decision model in the coexistence controversy

A different political practice of the Commission emerged with the coexistence controversy, a practice that in a sense returned to an earlier political pragmatism. But it is not the same political pragmatism encountered during the making of Directive 90/220. The coexistence controversy initiated a new kind of political pragmatism, more aware of its pragmatic nature and more sensitive in pursuing its political agenda, which could be described as a limited co-evolutionary decision model. It is limited in the sense that the 'framing assumption' can no longer be called into question to the same extent as before. Different assumptions about the socio-economic and political considerations are of course still there but they are not up for debate. The policy approach in the case of coexistence differed from the implementation of Directive 2001/18 in that the latter was concerned with establishing public confidence in the lawmaking process, and the former was concerned with implementing a new regulatory regime. It therefore needed less of a 'broad' consensus about the framing assumptions.

Millstone argues that, 'Framing assumptions influence not just which questions are asked, but also which types of evidence are included or discounted and how the selected evidence is interpreted' (ibid.: 500). Regarding the coexistence dispute, the concept of framing assumptions entails certain conditions for what to include and what to avoid in the agenda the Commission addresses, such as including a framework for coexistence but not including a framework for GMO-free regions. The framing assumptions define the types of questions that the Commission asks, and which to avoid or neglect; for example, including questions about the isolation distance to avoid mixtures but excluding liability issues. Another aspect is that these assumptions will define which type of evidence the Commission will deem relevant, and which they will ignore; for example, studies carried out by the EFSA are accepted, but the Müller study is neglected.[3] Furthermore, the concept of 'framing assumptions' would also explain how the European Commission considers scientific evidence and interprets them; for example, the Commission could have interpreted the Müller study as relevant but instead chose to consider it as not having provided any new scientific evidence as the sources referred to in the bibliography were published prior to the adoption of Directive 2001/18/EC. All these results are consequences of a policy making practice that entails a hybridisation of politics and science but also more profound structures of a historical, institutional, and economic nature.

Mapping the regimes of European GMO politics

The politics of coexistence indicated a return to a form of political pragmatism in the sense that coexistence framed GMO politics in scientific rather than political terms regarding safety zones, GM mixtures, and so on. The Commission's politics of coexistence displays a less reflexive approach by retreating back to the use of 'objective' scientific interpretations. The Commission has learned to steer in an extremely complex regulatory environment by choosing its battles much more carefully than before. For example, the earlier norm-setting authority in Directive 2001/18 was produced through a negotiation among a network of actors. However, in the case of coexistence it was also negotiated but in a more strategic sense. The Commission had chosen what to discuss within the regulatory framework of coexistence in a more active way by emphasising tolerance thresholds and distances between fields, which were to be settled by scientific expertise defined by the Commission. Also, letting the member states develop their own framework of coexistence brought them in as partners in norm-setting authorities, even if the boundaries that the member states had to operate within were unclear at the time.

Decisions on GM thresholds and specific isolation distances are pervaded by uncertainties. There are scientific uncertainties about the thresholds isolation distances for achieving a 0.9 percentage of GM mixture. The consequences of scientific uncertainties are then interpreted differently by the member states. In the politics of coexistence, the achieving of GM thresholds has caused new problems for institutionalised decision making, particularly when it comes to organic crops where the organic organisation desires a zero per cent threshold for GMOs – not possible in a European agricultural system that is based on GMOs. In order to limit the controversy over the coexistence framework, the Commission has withdrawn to a pragmatic form of politics and tried to bypass the controversies by returning to a more science-based system but was not able to restrict new regulatory conflicts.

Table 5.2 presents an overview of the regulatory regimes described in the first part of this chapter mapped on to the different policy models discussed in the second part. The case of developing GMO legislation in Europe illustrates that restricting the socio-economic and political considerations embedded in the regulation has not been possible. In that sense, the coexistence politics that the Commission endorsed for almost a decade has been unsuccessful. During this decade, the underlying framing assumptions have been problematic because there has been poor agreement on them. Finally, the Commission had to revise the regulations, allowing broader considerations for the member states to base their decisions on cultivating GM crops. The approach by the Commission to include a new paragraph in Directive 2001/18, and strengthen the member states autonomy for choosing to cultivate GM crops, may be seen as opening up the framing assumptions for discussion and allowing other normative, socio-economic, and political considerations to take place.

With the next generation of biotech innovations – which will utilise new biotech innovations in areas such as agri-food, agri-pharmaceutical and agri-industrial – there will also be new questions of scientific, normative, and political dimensions. It will be crucial for future regulatory regimes to consider seriously how the framing assumptions are structured in order to reduce future conflicts. The lesson learned from this study is that in a European setting there has to be a dialogue and shared understanding of the framing assumptions, meaning that the socio-economic and political considerations cannot be set by the European Commission and then be established through regulatory frameworks in the member states. Instead, to build stable regulatory frameworks that can deal with the new challenges in biotech innovations, these considerations have to be part of a shared project among the member states.

Table 5.2 Regimes of European GMO politics

Regimes of European GMO politics

Regulatory Regime Element	Directive 90/220	Directive 2001/18	Coexistence
Policy model	Inverted decisionism	Co-evolutionary	Limited co-evolutionary
Norm-setting authorities	Politics/policymakers and experts	Negotiated among a network of actors	Strategically negotiated among certain types of actors
Subject for regulation	Legitimate law and uncertainties with risks	Public confidence in the lawmaking process	Incorporating a new regulatory regime
Regulatory approaches	Addressing issues of social, political, and economic concerns within regulation	Increasing politicisation in the regulatory process	Decreasing politicisation in the regulatory process
Results	A new regulatory approach to GMO based on precaution, process, and horizontally-based	More consumer-sensitive regulation	New regulatory conflicts over different interpretations

Notes

* I would like to thank Daisy Laforce for comments on this chapter as well as participants at the Brocher Foundation Symposium 'Regulating next Generation Genomics: Emerging Agricultural Biotechnology Governance Challenges' on 8 July 2010.
1 The strategy for those scientists that could not agree with the dominant risk paradigm in the Asilomar conference turned to organisations that had not been represented at Asilomar. Examples of these organisations were Friends of the Earth and Science for the People. Through the organisation Science for the People, which consisted of scientists primarily from Cambridge and Boston, an open letter was sent to the participants at the conference stating that, 'Regardless of how slight the *current hazards* of introducing man-made hybrid microorganisms into the environment appear at the moment, the hazards of such activities are unknown and possibly great'. The letter is reprinted in Watson and Tooze 1981: 49 and is signed by a group of nine scientists from Harvard Medical School, Harvard University, Boston University, and MIT. However, Paul Berg's report from the Asilomar conference states that, '…it was recognised that future research and experience may show that many of the potential biohazards are less serious and/or less probable than we now suspect.' P. Berg *et al.* 'Potential Biohazards of Recombinant DNA Molecules', in Watson and Tooze 1981: 44.
2 GENET is a European network of non-governmental non-profit organisations engaged in the critical debate on genetic engineering, founded in 1995. http://www.genet-info. org/. (accessed 26/10/2008).
3 The study known as the 'Müller study' suggests that it is practically impossible for organic and conventional production to coexist with GMO farming without contamination. The study proposes a 4 kilometre protection zone as a minimum distance between GM cultivation and other cultivation areas. If one followed the recommendation, together with the fact that Austrian farming is small-structured, GM farming would not be possible at all (Müller 2002).

Bibliography

Berg, Paul, David Baltimore, Herbert W. Boyer, Stanley N. Cohen, Ronald W. Davis, David S. Hogness, Daniel Nathans, Richard Robin, James D. Watson, Sherman Weissman and Norton D. Zinder (1981) "Potential Biohazards of Recombinant DNA Molecules", in J. D. Watson and J. Tooze, *The DNA Story. A Documentary History of Gene Cloning,* W. H. Freeman and Company, San Francisco.

Council of European Union (1990a) *Directive 90/219/EEC of 23 April 1990 on the contained use of genetically modified microorganisms,* OJ L 117, 8.5.1990.

Council of European Union (1990b) *Directive 90/220/EEC of 23 April 1990 on the deliberate release into the environment of genetically modified organisms,* OJ L 117, 8.5.1990.

Council of European Union (2006) *Coexistence of genetically modified, conventional and organic crops – freedom of choice,* 9810/0, Brussels, 24 May 2006.

European Commission (1994) *Biotechnology and the White Paper on Growth, Competitiveness, Employment, preparing the next stage,* Communication from the Commission to the Council, the European Parliament, and the Economic and Social Committee, COM(94) 219.

European Commission (2000a) Communication from the Commission to the European Parliament pursuant to the second subparagraph of Article 251 (2) of the EC-Treaty concerning the Common Position adopted by the Council with a view to the adoption of a Directive of the European Parliament and of the Council on the deliberate release into the environment of genetically modified organisms and repealing Council Directive 90/220/EEC. COD (1992) 0072.

European Commission (2000b) *Communication from the Commission on the Precautionary Principle,* COM (2000) 1 Final (Brussels, 2 February).

European Commission. (2001) Directive 2001/18/EC of the European Parliament and of the Council of 12 March 2001 on the deliberate release into the environment of GMOs and repealing Council Directive 90/220/EEC [2001] L106/1.

European Commission (2002) *Life sciences and biotechnology – A Strategy for Europe,* Communication from the Commission to the Council, the European Parliament, the Economic and Social Committee and the Committee of the Regions of 23 January 2002, COM(2002) 27 final, (O. J. 2002/C 55/03).

European Commission (2003) *Recommendation on guidelines for the development of national strategies and best practices to ensure the coexistence of genetically modified crops with conventional and organic farming.* OJ L 189/3.

European Commission (2009) *Report from the Commission to the Council and the European Parliament on the coexistence of genetically modified crops with conventional and organic farming.* COM(2009) 153 final.

European Commission (2010a) *Proposal for a Regulation of the European Parliament and of the Council amending Directive 2001/18/EC as regards the possibility for the Member States to restrict or prohibit the cultivation of GMOs in their territory,* COM(2010) 375 final, Brussels.

European Commission (2010b) *Communication from the Commission to the European Parliament, the Council, the Economic and Social Committee and the Committee of the Regions on the freedom for Member States to decide on the cultivation of genetically modified crops,* COM(2010) 380 final, Brussels.

European Council (2001) *Directive 2001/18/EC of the European Parliament and of the Council of 12 March 2001 on the deliberate release into the environment of GMOs and repealing Council Directive 90/220/EEC,* OJ L 106, 17.4.2001.

European Parliament (1997) Regulation 258/97/EC of the European Parliament and of the Council (OJ L43, p1, 14/02/1997) of 27 January 1997 concerning novel foods and novel food ingredients amended by Regulation 1829/2003 of the European Parliament and of the Council (OJ L268, p1, 18/10/2003) of 22 September 2003 on genetically modified food and feed.

Fredrickson, D. S. (2001) *The Recombinant DNA controversy. A Memoir – Science, Politics, and the Public Interest 1974–1981*. Washington DC: ASM Press.

GMO-free-Regions.org (2005) 'Berlin Manifesto for GMO-free Regions and Biodiversity in Europe – Our Land, our Future, our Europe', http://www.gmo-free-regions.org/conference-2005/berlin-manifesto.html (accessed 26/10/2008).

Gottweis, H. (1998) *Governing Molecules: The Discursive Politics of Genetic Engineering in Europe and the U.S.* Cambridge: MIT Press.

Johansson, A. (2009) *Biopolitics and Reflexivity: A Study of GMO Policy Making in the European Union*, PhD diss., Linköping Studies in Arts and Science, Linköping universitet, Linköping.

Krimsky, S. (1982) *Genetic Alchemy*. Cambridge: MIT Press.

Millstone, E. (2007) 'Can food safety policy-making be both scientifically and democratically legitimated? If so, How?', *Journal of Agriculture and Environmental Ethics* 20, no 5: 487–8.

Müller, W. *Genetically modified-free areas of farming: conception and analysis of scenarios and steps for realisation*, Commissioned by the Department for the Environment of the Region Upper Austria and the Federal Ministry for Social Security and Generations. (28 April 2002).

OECD (1986) *Recombinant DNA Safety Considerations*.

Shore C. and Wright S. (1997) *Anthropology of Policy. Critical Perspectives on Governance and Power*. Routledge, London.

Watson, J. D. and Tooze, J. (1981) *The DNA Story. A Documentary History of Gene Cloning*. San Francisco: W. H. Freeman and Company.

Wright, S. (1994) *Molecular Politics: Developing American and British Regulatory Policy for Genetic Engineering, 1972–1982*. Chicago: University of Chicago Press.

Wynne, B. (1996) 'May the sheep safely graze? A reflexive view of the expert-lay knowledge divide', in Lash, Szerszynski and Wynne (eds) *Risk, Environment and Modernity. Towards a New Ecology*. London: SAGE Publications.

Zwanenburg P. Van, and Millstone, E. (2005) *BSE: Risk, Science, and Governance*. Oxford: Oxford University Press.

Part 3

GMO regulatory regimes in practice: Europe, Asia, and North America

6 Contested frames

Comparing EU versus US GMO policy

Sarah Lieberman and Anthony R. Zito

Introduction

In 2003, the USA, Canada, and Argentina lodged a World Trade Organization (WTO) complaint against the EU's moratorium on authorisations of new genetically modified (GM) crops and foods. This led to the longest ever WTO dispute settlement – the Panel did not release its findings until 2006. The Panel concluded positively for the complainants, finding the EU guilty of 'undue delays' in the approval process for biotech products (WTO 2006d). Nevertheless, the Panel did not issue sanctions against the EU – the EU was simply told to ensure in the future that its authorisation procedure adhered to its own rules and regulations. Four years on, GM foods are still not readily available within the member states of the European Union; indeed, sourcing GM products in shops and supermarkets is difficult. Whereas in the USA, the opposite remains true: non-GM varieties of certain products are scarce. Therefore, the positive outcome of the WTO challenge appears to have had very little impact on the availability of GM foods and crops in the EU, and therefore little impact on the USA's export market for the products of agricultural biotechnology.

This difference in GM food availability illustrates a wider historical divergence between the two regulatory systems. In the EU, GMOs (including GM crops and foods) are regulated automatically because 'being genetically modified' demands that certain procedures, including a stringent authorisation process, are followed. In the USA, GM products are treated in the same way as conventional produce. This does not mean that they are '*unregulated*'; rather, they must conform to exactly the same safety standards as non-GM seeds, non-GM plants, non-GM pesticides, and non-GM foods. This difference is often described as the difference between a process oriented regulatory framework in the EU (where the process of development through genetic modification means that GM products are treated separately), and a product-oriented framework in the USA (where the product, regardless of development technique, must adhere to certain standards).

However, the regulatory difference is not merely procedural. Strong ideational factors have been present in the debate over GMOs since the outset of the technology, and in both the EU and the USA, the rules, regulations, and procedures have become strongly institutionalised. Indeed, the international debates over regulatory standards and frameworks mask a deeply entrenched divergence in the ideational and normative debate regarding 'likeness': the process oriented framework of the EU suggests a riskiness to GM products unseen in conventional agricultural produce. GMOs are therefore not seen as 'like' conventional products. The product-oriented framework of the USA suggests that GMOs are no more risky than the products of conventional agriculture, and are thus 'like' conventional products. Consequently, we have two opposing frameworks of regulation for agricultural biotechnology and its products, conflicting on procedural, ideational, and institutional grounds.

How the divergence between the EU and the USA is characterised differs between commentators: the distinction between product and process oriented approval processes is highlighted by Dunlop (2000) and Jasanoff (1995); the distinction between substantial equivalence and the precautionary principle is highlighted by Lieberman and Gray (2006); and Bernauer (2003) notes a difference of promotion and restriction. Moreover, Bernauer also highlights a difference between 'bottom up' regulation in the EU (where a level of new governance input from a wide range of actors has characterised the development of policy) and 'top down' regulation in the USA (where federal governmental control dominates the regulatory framework). Most interesting is that all of these divergences exist, and often without precedent. As Wiener and Rogers (2002) note, we cannot state that the EU is always more dependent on the precautionary principle than the USA – we need only look at the continuing trade issue of BSE for confirmation; and similarly we cannot state that the USA is always keener to promote new technologies than the EU.

To explain this divergence, this chapter emphasises the nature of framing at the heart of the regulatory system. Majone (1996: 9) defines regulation as a set of 'rules issued for the purpose of controlling the manner in which private and public enterprises conduct their operations', but this regulatory activity may avail itself of several different modes. He differentiates between the type of regulation traditional to the United States (where a public agency exercises control on the basis of a legislative mandate over activities held to be desirable in society), and to Europe, which has responded to market failures by having more intrusive forms of market control (particularly public ownership and *dirigisme* – see Majone 1996: 9–23). He and other scholars of regulation have argued that, since the 1980s, Europe has seen a shift in regulation towards independent agencies, rule-making, experts, and courts instead of political parties and civil servants – the 'regulatory state hypothesis' (Thatcher 2002: 867).

Such regulatory analysis suggests that we must look at the key actors and processes behind the regulation, but we argue further that the *framing* of the key policy problem (that is the objective of the regulatory response) may shape some of the larger macro choices. Derived from public policy and social movement theory, 'policy framing' is a concept seeking to explain the process by which social and political actors understand and shape complex situations (Goffman 1974; Rein and Schön 1993). It focuses on how actors conceptualise the world, how the elements of the world interact, and how policy actors position themselves. The emphasis is on the dynamic of change and the constantly changing and contested notion of policy reality (Schön and Rein 1994).

Public policy will be shaped by the contestation between frames where persuasion about one's perspective, rather than a vindication of truth, determines the outcomes. This contestation involves a policy dialogue, or discourse, between those actors involved in any given policy issue. Conflicting policy frames mobilise policy actors to fight the status quo. A process of 'reframing' occurs when the reflection upon the frames and the discourse leads to a resolution of the policy controversies. The intersubjective engagement with other actors and their own personal situation will lead the actors to identify incongruities and changes in the particular environment, triggering frame change.

Schön and Rein (1994: 31–6) further differentiate policy discourse into the following elements. The first discourse centres around policy actors seeking to define the policy debate getting their policy stories/frames accepted as dominant. Rhetorical frames will underpin the arguments in the policy debate. The second discourse describes the policy debate involving the actors creating and then contesting policy stories that inform both the policy procedures and particular policy instruments that engage the policy. It is this second discourse where the discussions of particular modes of regulation and governance will be fought.

Further complicating the divergence in regulatory frameworks is the strong global governance input on this issue. National, federal, and supranational governments must adhere to the rules, norms, and institutional guidelines provided by international institutions including the WTO, the Convention on Biodiversity (CBD), and the World Health Organization (WHO). Additionally, there has been a strong 'new governance' aspect to the debate; civil society reaction to GMOs, on both sides of the Atlantic has been strong, with environmental groups, consumer groups, industry groups, and 'sound science' all contributing to the development of the regulatory framework. However, despite input and interest from other parties in both the USA and the EU, the regulatory framework for GMOs has been largely driven by traditional government structures.

The argument

This chapter investigates the differences between the EU and the USA in the wake of the WTO challenge. First, what is the basis of the regulatory differences between the two? To discuss this issue, the chapter assesses the differences in policy frames within a historical overview of GMO regulation in the EU and the USA, outlining the basis and outcomes of the WTO 'biotech products' challenge. Second, what has actually changed as a result of the ruling issued by the WTO dispute settlement panel? This chapter will provide an analysis of the current situation to conclude whether or not there has been any change to GMO regulation in either the EU or the USA, following the WTO challenge.

EU policy history

Since the outset of biotechnology regulation in the 1980s, the European Union has followed its own path, diverging early on from the path followed by the

United States. This has led to what is often referred to as a globally polarised regulatory arena, in which the European Union follows a stringent, precautionary, environmentally based, restrictive legislative framework and the United States uses existing regulation to ensure public safety through a 'permissive' legislative framework. It is often argued that this polarised system has provided a choice of legislative trajectories for other countries, particularly developing countries (Falkner and Gupta 2009: 114), and that the two major global governance regimes for biotechnology (the CBD's Biosafety Protocol and the WTO), are based on the principles established by the EU and the USA respectively. In this section, we will examine the history of the EU's regulatory framework for GMOs and the framing principles (process, precaution) that underpin it, as well as how this framing interacted with the dynamics of path dependency and public opinion.

Process oriented

The EU has a 'process' based system of regulation for GMOs (the framing of the nature of the problem, as well as the policy solution and policy goal). The EU has made a critical choice to focus on the perspective that as GMO products are being manufactured using biotechnology *processes*, each product must undergo stringent testing, authorisation procedures, and post market monitoring. Directives 90/220 on the Deliberate Release, and 90/219 on the Contained Use of GMOs came into force in 1991 following much discussion and debate. Directive 90/220, in particular, is seen as a pivotal piece of legislation, and the process based authorisation procedure it defined has remained central to the legislative framework for agricultural biotechnology. Directive 90/220 was split into two sections; part B for small scale research trials and part C for commercial releases. Part B specified that GMOs should be subject to a 'case-by-case' environmental risk assessment and a 'step-by-step' process should be applied to any release deemed necessary for research purposes: that is, tests would start on a small scale and increase only if judged safe. Under 90/220, Part B research requests were dealt with, and directly controlled by the national 'competent authority' in the member state in which the field trials or research would take place.

According to the preamble of Directive 90/220, the harmonisation of GMO legislation in Europe under its terms is necessary to prevent 'unequal conditions of competition or barriers to trade in products containing such organisms [GMOs] thus affecting the functioning of common market' (CEC 1990). Despite being an early piece of legislation, Directive 90/220 was thus wide ranging – it covered environmental and economic concerns and the harmonisation of member states' laws concerning GMOs. However, it came under increased criticism as biotechnology progressed and Europe suffered several food scares during the 1990s.

In 1996, a decision was taken to strengthen the regulatory framework for GMOs (Europa 1996); in October 2002 Directive 90/220 was repealed by Directive 2001/18; in September 2003 the Novel Foods Regulation was replaced by the more stringent Regulation 1829/2003 on Genetically Modified Food and Feed; and stricter rules concerning the labelling and traceability of GM food and feed were also added to the regulatory framework in September 2003 by Regulation 1830/2003. However, on account of the slow moving nature of legislative drafting, in June 1999 member states issued declarations stating their intent to block all applications for authorisations of new GMOs until the new rules on GMO approvals could take effect (Marris 2000). This meant that a de facto moratorium was applied to all new GMO authorisations.

Due to the de facto nature of the moratorium, its terms and conditions were not entirely clear. However, it now seems that two moratoria were in fact applied: one on new authorisations of GM foods and one on new authorisations of GM crops. The first (on foods) ostensibly lasted from 1999 (when the moratorium declarations were signed) until 2003 when the *Biotech Products* WTO case was launched; yet during this time frame some GM foods and food derivatives were authorised for sale and consumption (Lieberman and Gray 2006: 605). The second (on crops) lasted from June 1999 until 2010 when the first GM crop since 1998 was approved for cultivation. The Amflora potato was authorised for cultivation for industrial uses in March 2010 (CEU 2010).

The new Deliberate Release Directive extended the precautionary requirements of the notification procedure. Under Article 13:2, all notifications must contain an environmental risk assessment; a proposed period for the consent which must not exceed ten years; a plan for monitoring, including a time period for the monitoring plan; and proposals for labelling and packaging (CEU 2001).

Precautionary principle

The EU is committed to using the precautionary principle to protect the environment, and the case of GMO regulation has been its greatest test so far. The precautionary principle was first outlined in the UN's World Charter for Nature in 1982 which states that, 'activities which are likely to pose a significant risk to nature shall be preceded by an exhaustive examination; their proponents shall demonstrate that expected benefits outweigh potential damage to nature, and where potential adverse effects are not fully understood, the activities should not proceed' (UN 1982). By the values it embodies, this principle reinforces a framing of GM products emphasising issues of risk, uncertainty, and potential damage. In 2000, the European Commission released a Communication stating how the precautionary principle should be used, even though it had been present as a strong and ever strengthening element of GMO regulation since the outset more than a decade earlier. The reason for the Commission's clarification is the ongoing debate regarding the basis, applicability, and legality of the precautionary principle. As the Communication states, '(T)he issue of when and how to use the precautionary principle, both within the European Union and internationally, is giving rise to much debate, and to mixed, and sometimes contradictory views' (CEC 2000).

Many commentators see the use of the precautionary principle as the major dividing factor between the USA and the EU in terms of GMO regulation (i.e. Geuhlstorf and Hallstrom 2005). However, Lieberman and Gray (2008a: 396) suggest that both the EU and the USA apply aspects of precaution in relation to the regulation of GMOs – both new crops and foods must undergo a process of approval. The difference is that the EU assumes 'potential difference' between GM crops and conventional crops while the USA assumes 'substantial equivalence' between GM and conventional crops and foods in their regulatory framework. Cantley (1995) notes that, until 1986, when the EU began drafting the GMO directives, substantial equivalence was considered an adequate principle through which to regulate the newly emerging agricultural biotechnologies. This means that a product considered 'substantially equivalent' to a known product can be authorised using the same legislative framework as the known product. Aspects of this principle survived the drafting of European regulation for GMOs. The original pieces of legislation on GMOs, the Deliberate Release (90/220) and the Novel Foods and Novel Food Ingredients Regulation (258/97), both included aspects of substantial equivalence in their simplified authorisation procedures (Lieberman and Gray 2008a: 398).

However, the revised Directives and Regulations on GMOs repealed these procedures, strengthening the EU's precautionary approach to agricultural biotechnology and removing the principle of substantial equivalence from the legislative framework for GMOs. This policy history tells us that the precautionary principle has been a part of GMO regulation for as long as GMOs have been regulated in the EU, but that 'precaution' has been an ever strengthening position for Europe, as well as the initial legislative policy position. This reinforces the argument that there has been an underlying framing of the policy problem that is independent to generic policy principles, which can be articulated by various actors to signify different things.

Path dependency

The institutional structure of the EU creates a strong degree of path dependency. This is particularly clear on the issue of GMOs. The choices made early on, including (but not limited to) the framing of agricultural biotechnology as an environmental issue, has had probably the biggest impact on the development of the legislative framework for GMOs.

The underlying principles of the EU's regulatory framework were established in the mid-1980s when the EU broke away from the USA in terms of outlook regarding the emerging biotechnologies. Given the level of negative press attention and negative public opinion on the issue of agricultural biotechnology since then, we may be given to assume that the EU's GMO regulatory structure was framed as an environmental issue, rather than an agricultural development or scientific breakthrough, because of societal pressure. However, we note that this policy area was framed as environmental by the European institutions before any real public or media interest developed on the issue. In 1985, DG (Directorate General) XI (Environment) was charged with the task of co-chairing BRIC (the Biotechnology Regulation Interservice Committee) with DG III (Internal Market and Industrial Affairs) to handle the technical side of regulation development for the Biotechnology Steering Committee.

As co-chairs of BRIC, DG Environment and DG Internal Market and Industrial Affairs were given responsibility to draft the EU's GMO regulatory policy. The decision had to be taken at this time as to whether the process of gene splicing was an activity that required oversight similar to that required for 'hazardous industrial activities', and whether the resultant products would automatically require regulation similar to that developed by DG Environment for chemicals. DG Environment did the bulk of this work, according to Patterson (1997: 30), 'because DG III was too busy to participate in either the drafting sessions or the Council Meetings'. By allowing DG Environment to draft the proposals, and therefore present their work to the Council of Environment Ministers, biotechnology was framed early on as an environmental issue to be regulated in a manner similar to other environmental release issues and treated with the same level of control and 'harmonised Community-wide procedures' (Cantley 1995: 547).

Institutionally, the issue was framed early on as environmental. DG Environment was appointed *chef de file* under Directive 90/220 on the Deliberate Release into the Environment of GMOs, and remained *chef de file* for all Deliberate Release authorisation requests lodged under Directive 2001/18 until February 2010 (CEU 2010a). Procedurally, as Falkner and Gupta (2009: 118) note, the directives 'institutionalised a process based approach with comprehensive coverage of all GMO developments' and, in terms of risk strategy, Directives 90/220 and 90/219 were strongly based on the precautionary principle which 'legitimises regulatory intervention to avert potentially serious or irreversible harm under conditions of scientific uncertainty' (Falkner and Gupta 2009: 118).

USA policy history

In the USA, the bulk of certain agricultural crops (including soybeans, maize/corn, cotton, and rapeseed) are grown using GM varieties. Strauss (2008: 779) notes that 'in 2007, ninety-one percent of all soybeans, eighty-seven percent of cotton, and seventy-three percent of corn acres were planted with GE varieties, mainly to control weeds and insects'. The political discourse of 'likeness' between GM and conventional crops in the USA means that agricultural biotechnology is *framed* as a tool of industrial agriculture like any other and, as such, GM and conventional crops are not segregated (Strauss 2008: 775). This means that a very high percentage of processed foods available in the USA contain GM ingredients. Although exact figures are almost impossible to estimate, Ackerman (2002) suggests that in 2002, sixty per cent of all processed foods on US supermarket shelves contained GM ingredients.

Regulatory strategy

In the USA, biotechnology is regulated at the federal level. Although there have been attempts to implement county level GMO bans in certain states (i.e. California, Vermont, and Oregon), these have not seen much success, and the regulatory system has remained stable – strongly federal and centralised.[1] Since 1986, agricultural biotechnology has been regulated under the 'coordinated framework' of two federal agencies and one federal department: the United States Environmental Protection Agency (EPA), the United States Food and Drug Administration (FDA) and the United States Department of Agriculture (USDA).

The framing decision was made early in the history of the technology, that existing legislation and existing government departments were best placed to ensure public and environmental safety and that GMOs should be 'handled within the regulatory framework that existed prior to their invention according to their characteristics, such as nutritional content, rather than their means of production' (Strauss 2008: 781). As such, the coordinated framework states that '(E)xisting statutes provide a basic network of agency jurisdiction over both research and products; this network forms the basis of this coordinated framework and helps assure reasonable safeguards for the public' (Office of Science and Technology Policy 1986). Agricultural biotechnology is not unregulated in the USA, but there has never been legislation written specifically for GMOs. At the outset, this was justified in the coordinated framework document as providing 'more immediate regulatory protection and certainty for the industry than possible with the implementation of new legislation' (Office of Science and Technology Policy 1986: 4).

The coordinated framework (and its underlying framing) is product based and functions vertically. Although the responsibilities of the three agencies are 'complementary, and in some cases overlapping' (US Biotech Reg. 2010), each has responsibility for different classes of product; broadly, pesticides, foods, and plants. The EPA uses pre-existing policies to regulate 'pesticidal and nonpesticidal microorganisms' (Office of Science and Technology Policy 1986: 4), meaning it has responsibility for those products, plants, and the resulting food and feed, modified to contain a pesticide. The FDA regulates GM foods and feed in the same way, and to the same standard, as traditional, conventional products. The USDA has responsibility for all GM animals and the planting of GM crops and plants.

Because of the regulatory focus on the safety of the end product, and the use of existing policy and departments to assess this, the regulatory strategy for GMOs in the USA is often said to be 'product' based, that is 'based on the assumption that there is nothing unique or particularly risky in applying genetic engineering in agriculture, and that products resulting from such technology are usually the same as comparable products obtained via conventional agricultural methods' (Bernauer 2003: 44), as opposed to the EU's process based system. Indeed, as Strauss (2008: 781) notes, the process of product development is largely ignored: 'the Food and Drug Administration (FDA) does not recognise any inherent risk in, or indeed any relevance of, the source of the product'.

These framing movements made in the mid-1980s have become institutionalised, and although it was suggested in the coordinated framework that legislation would change with the evolving technology (Office of Science and Technology Policy 1986: 4), the regulatory strategy has remained remarkably stable. This suggests that agri-biotech regulation, and the framing that underpins it, follows a strongly path dependent trajectory.

Substantial equivalence

The manner of regulating GM crops and foods in the USA depends on their acceptance as being 'like' conventional products. This means that the regulatory framework is based on the principle of substantial equivalence. This is a difficult principle to define and it has come under criticism for its lack of precision. For example, Michael Meacher states that 'science is not about saying that there is A and B and they are substantially equivalent: they are either equivalent or they are not, and if they are not equivalent we want to know in exactly what respects they are not the same, and the percentage difference – even if it is only 0.00001%' (Meacher 2004). However, as Miller (1999: 1042) notes, 'substantial equivalence is not intended to be a scientific formulation; it is a conceptual tool for food producers and government regulators, and it neither specifies nor limits the kind or amount of testing needed for new foods'. Indeed, despite its critics, substantial equivalence is an important regulatory tool, and, as noted by Lieberman and Gray (2008a: 397), 'it is used as a benchmark for the assessment of GM food safety by the OECD and the Food and Agriculture Organisation (FAO)'.

Strong federal government control

Although commentators often suggest that GMOs are *not* regulated in the USA, this is not the case. Agricultural biotechnology and its products are subject to strictly centralised federal oversight. In the USA, individual states must abide by federal agricultural rules. As one Animal and Plant Health Inspection Service (APHIS) official explains, all agricultural products, including those developed using agri-biotech techniques must be regulated at the federal level: 'when the feds regulate something, they solely regulate it, and this is true of this particular question because it involves interstate commerce and agricultural practice. Produce needs to freely move through all borders within the country and if individual states were to put restrictions on that it would harm interstate commerce' (Wach 2005). Bernauer (2003: 111) describes this permissive but strongly federal system of regulation as 'centralised laxity', as opposed to the ratcheting-up regulatory system for GMOs seen in the EU.

However, if we look at other aspects of US environmental policy, centralisation does not preclude increased standards. As Vogel (1997: 561) notes, the term 'California Effect' whereby 'national environmental regulations…[move] in the direction of political jurisdictions with stricter regulatory standards' was coined following increased automobile emission standards in the USA based on California's stricter controls – a ratcheting-up process. So why has this process not occurred in the USA's regulatory system for GMOs? Bernauer (2003) notes both public opinion and governmental explanations for this. First, he notes an absence of the 'bottom up' public pressure against GMOs seen in Europe. Second, he points out that, on the issue of agri-biotechnology, US states do not have the same level of autonomy as EU member states, and thus do not have the same 'formal opportunities…to ratchet up federal regulatory stringency through majority decision making' (Bernauer 2003: 114).

But this does not explain why GM policy differs from previous regulatory strategies in the USA. Vogel (2003) addresses this by adding a chronological explanation to his analysis. Vogel (2003: 563) notes that on GMOs public opinion and governmental strategy are important: 'American regulatory officials have worked cooperatively with industry to facilitate the commercial development of this new technology. There has been relatively little public participation in the regulatory process and only intermittent public scrutiny of regulatory decisions'. Vogel also suggests (2003: 578) that, due to consumer health and environmental disasters in the EU during the 1980s, regulatory strategy has been tightened, while an absence of environmental and consumer policy failures in the USA during the same time frame means that regulatory strategy has loosened. He draws a parallel between environmental policy in the USA from 1960 to 1990 and the EU since the mid-1980s, using supersonic air travel in the USA and GMOs in the EU to illustrate a similarity across the twenty-five year gap. '[I]n both cases, a significant segment of the public saw no benefits associated with the proposed new technology, only increased environmental and health risks' (Vogel 2003: 576). This argument suggests that the general acceptance of policy frames for environmental issues may be reliant on the evolving attitude of the mass public towards questions of risk and new technologies.

The WTO arena: biotech products

In 2003, the USA, Canada, and Argentina launched a WTO challenge against the EU regarding its lack of biotech authorisations. According to the submitted documents, the challenge aimed to end a perceived three pronged moratorium on authorisations – a general moratorium on licensing, a moratorium on specific products, and member state bans on GMOs (WTO 2006d). In short, by lodging a challenge at the WTO, the complainants sought to remove those aspects of EU GMO policy obstructing free trade between the EU and the USA, Canada, and Argentina in GM food and crops and to end the general moratorium on new GM crops and foods. The initial complaints issued refer to Article 11 of the SPS (sanitary and phytosanitary) Agreement and Article 14 of the Technical Barriers to Trade Agreement. These articles specify the mobilisation of the dispute settlement procedures of the World Trade Organization (WTO 2006d), suggesting that the complainants believed the EU's GM authorisation procedure contravened the terms of the EU's WTO obligations under different agreements.

In its defence, the EU claimed it had not applied a moratorium on GM products. This claim was denied by the dispute settlement panel who decided that the moratorium had indeed existed. However, the Panel ruled that the moratorium was not a 'measure' as defined by the SPS Agreement, and therefore could not contravene several of its obligations. Moreover, the Panel felt unable to rule on whether biotech products should be considered 'like' their conventional counterparts (WTO 2006d). This meant that they could not rule under the Technical Barriers to Trade Agreement, which ensures that there is no trade discrimination between like products. Accordingly, the Panel took a stance in which it refused to assess the framing of the GMO issue by the EU.

Although a victory was noted for the complainants, the outcomes are interesting. As Lieberman and Gray (2008b: 33) note, the 'win' was on a relatively narrow basis. The EU was found guilty only of contravening 'one clause, of one section, of one annex' (Lieberman and Gray 2008b: 45) with its failure to complete authorisations in a timely fashion. Importantly, several of the issues that were brought to the attention of the WTO dispute panel were not commented upon: specifically, whether GM products can be considered 'like' conventional products and therefore whether or not the moratorium on GMO authorisations constituted a technical barrier to trade. Moreover, the dispute settlement panel ruled that the level of stringency applied by the EU to GMO authorisation was justifiable and that under some circumstances a moratorium on GMOs would be justifiable. And so the contrasting framing of the GMO issues remains in the international arena, and is no closer to being resolved.

Outcomes

The WTO *Biotech Products* challenge illustrates not only issues regarding the trade in GMOs, but also the global regime for the regulation of agricultural biotechnology. Falkner and Gupta (2009: 118) note that '(B)oth the US and the EU have sought to export their regulatory models to the international level as well as to other countries'. On this point, they note that the US model and the EU model are echoed by the WTO and the CBD Biosafety Protocol respectively. The WTO's role is to maintain free trade and to prevent trade discrimination and protectionism on a global level, and its dispute settlement principles are designed to be 'equitable, fast, effective [and] mutually acceptable' (WTO 2006a).

In its bid to ensure that global trade flows 'as freely as possible', the WTO encourages and enforces mutual recognition with penalties and a programme to make certain 'that individuals, companies and governments know what the trade rules are around the world' (WTO 2006b). The WTO does not set standards, but works to ensure that markets interact without showing preference or prejudice. As cases are resolved, this can mean either a 'ratcheting-up' of global environmental standards if key markets have high standards; or a lowering of global environmental policy standards if key markets demand free trade over environmental policy issues. However, specific environmental agreements are only considered relevant in the eyes of the WTO if all parties to a dispute have ratified its terms; i.e. if 'one side...has not signed the environment agreement' the WTO assigns itself as arbitrator to the dispute (WTO 2006c). In practice, this often means that the WTO enforces environmental standards that are lower than those applied by some parties to disputes.

The Biosafety Protocol of the Convention on Biodiversity (CBD) has a much narrower remit and stipulates common standards for GM products: 'The Parties shall ensure that the development, handling, transport, use, transfer and release of any living modified organisms are undertaken in a manner that prevents or reduces the risks to biological diversity, taking also into account risks to human health' (CBD 2000). Although these standards are less stringent than hoped for by many, they are more stringent than others would wish; in other words, the Biosafety Protocol has aimed to achieve a compromise form of harmonisation. However, although the preamble of the Biosafety Protocol states that it should not be considered subordinate to any other international agreement (CBD 2000), the WTO's dispute settlement panel could not take it into account as it has not been signed or ratified by the USA. By issuing a challenge to the EU through the WTO, the USA therefore appealed not only to the regime closest to its own ideological discourse on agricultural biotechnology and closest to its regulatory framing strategy for GMOs, but also challenged the legitimacy of the only specific global regime for GMOs.

Both the EU and the USA have formulated their own policy frameworks for the regulation of agricultural biotechnology and its products to ensure that interstate trade is not hindered. However, as Bernauer (2003) notes, this has been done very differently by each. As already mentioned, in the USA, he notes a strong federal system of 'centralised laxity'; in the EU he notes a harmonisation system of 'ratcheting-up'. There are, therefore, two potential ways in which agri-biotech regulation could be harmonised to ease trade between the USA and the EU in GM products. First, the USA (as the exporting state) could comply with EU standards and undergo a process of essential harmonisation or 'forgo the benefits of export' (Trachtman 2007: 784). Second, the EU could accept the USA's lower level of regulation and accept imports from the USA. The second option is obviously the optimum outcome for the challengers, and the launch of a dispute settlement challenge suggests a preference for regulation enforced by the WTO.

However, as Vogel (1997: 556) notes, global economic interdependence does not naturally lead to 'a weakening of national environmental standards' and this is clear in the *Biotech Products* conclusions. Despite the WTO ruling against the EU's moratorium, no action was taken against the EU's GMO regulatory strategy and no sanctions were issued. Assessing the outcome, Lieberman and Gray (2008b: 59) suggest that the case was 'a pyrrhic victory for the USA' as it did not solve many of the underlying divergences regarding the precautionary principle, substantial equivalence and the framing question of likeness. The EU was asked to ensure that long delays on authorisations of GM products ceased, and certainly, some GM products have been approved since the biotech products case concluded in 2006. However, the first product for cultivation only gained approval in March 2010 – four years after the publication of the biotech products dispute panel settlement. Can it truly be said that the approval process (without any undue delay) takes four years to complete? Nonetheless, there is evidence that the EU's approval process is again moving: more products have been proposed, several have gained approval for sale and consumption under the food and feed regulation, and one has gained approval for cultivation for industrial use.

Underpinning the debate is the wider issue regarding US and EU trade with third parties, and on this issue it will be important which of the two models for GM regulation will prevail. On the one hand, it looks like the EU regulatory model is winning. The 'prisoner's dilemma' model suggests that the presence of a strong regulatory process will increase global levels of stringency:

Because both states find themselves in an integrated market they cannot decide in isolation. If both states pursued the policies they like most the common agricultural market would be disrupted...If the pro-agri-biotech state stuck to lax regulation, the anti-biotech state would, under pressure from voters (consumers), adopt strict regulation and risk a breakdown of the common market rather than agree to downward harmonisation. Once the anti-agri-biotech state opts for strict regulation and credibly signals to the other state that it will not back down, the pro-agri-biotech state has a strong incentive to adjust its regulation upward (Bernauer 2003: 106).

Moreover, globally there is a preference for ratcheting the pressure up, rather than downward. There has been the refusal of GM food aid in some African countries; a rejection of agricultural biotechnological advances in European member states and in several developing countries; widespread consumer preference for GM-free products; and governmental support for a precautionary approach to technology. However, the USA is very powerful in terms of trade, economic and political influence, and so its (de)regulatory model for GM technology may in the end win by default, or through insistent pressure and technological transfer. However, if the current situation persists and there is no regulatory strengthening in the USA and no regulatory loosening in the EU, the polarised regulatory framework may be here for the foreseeable future.

This international dispute confirms the insights suggested by the policy framing literature. It is extremely difficult on issues of complexity and uncertainty to reconcile two contrasting sets of policy frames and values. The dispute then comes down to the EU and the US seeking to persuade each other and third parties of the merit of their positions, without having the 'facts' to falsify each other's position. In the absence of increased knowledge of the effects of GMO products, the political resources and strategy that each side deploys may determine the winner.

Note

1 Bernauer (2003: 112) notes that, of the 158 pieces of GMO legislation submitted in thirty-nine states in the 2001/2002 period, only forty-five were adopted. Of these, none imposed any restrictions on the cultivation of GM crops or labelling requirements on GMOs.

Bibliography

Ackerman, J. (2002) 'Science and Space', *National Geographic Magazine*, May 2002, available at (http://science.nationalgeographic.com/science/article/food-how-altered.html, accessed 3 February 2010).

Bernauer, T. (2003) *Genes, Trade and Regulation: The Seeds of Conflict in Food Biotechnology*. Oxford: Princeton University Press.

Cantley, M. (1995) 'The Regulation of Modern Biotechnology: a Historical and European Perspective', in Dieter Brauer (ed.), *Biotechnology* (second edition) *Vol. 12: Legal, Economic and Ethical Dimensions*. New York: VCH.

CBD (2000) *Text of the Cartagena Protocol on Biosafety*, Convention on Biological Diversity 29 January 2000, available at (http://bch.cbd.int/protocol/text/, accessed 1 February 2011).

CEC, Commission of the European Communities (1990) 'Council Directive of 23 April 1990 on the Deliberate Release into the Environment of Genetically Modified Organisms' (90/220 EEC) Belgian Biosafety Server, (http://www.biosafety.be/GB/Dir. Eur.GB/Del.Ref/90.220.html, accessed 11 February 2005).

CEC, Commission of the European Communities (2000) 'Communication from the Commission on the Precautionary Principle', (http://ec.europa.eu/dgs/health_consumer/library/pub/pub07_en.pdf).

CEU, Commission of the European Union (2001) 'Directive 2001/18/EC of the European Parliament and of the Council on the deliberate release into the environment of genetically modified organisms and repealing Council Directive 90/220/EEC', *Official Journal of the European Communities* 17.4.2001: Page no. L 106/1 (http://europa.eu.int/eur-lex/pri/en/oj/dat/2001/l_106/l_10620010417en00010038.pdf , accessed 11 February 2005).

CEU, Commission of the European Union (2010a) *Minutes of the 1095 the meeting of the Commission Held in Brussels (Berlaymont) on Wednesday 17 February 2010 (morning)* PV (2010) 1905 Final 3 March 2010, Available at (http://ec.europa.eu/transparency/regdoc/rep/10061/2010/EN/10061-2010-1905-EN-F-0.Pdf, accessed 26 October 2011).

CEU, Commission of the European Union (2010b) *Commission Decision (2010/135/EU): concerning the placing on the market, in accordance with Directive 2001/18/EC of the European Parliament and of the Council, of a potato product genetically modified for enhanced content of the anylopectin component of starch* 2 March 2010, Available at: (http://eur-lex.europa.eu/LexUriServ/LexUriServ.do?uri=OJ:L:2010:053:0011:0014:EN:PDF, accessed 25 October 2010).

Dunlop, C. (2000) 'GMOs and Regulatory Styles', *Environmental Politics*, 9 (2): 149–55.

Europa Press Room (1996) 'Commission presents Report on Directive 90/220/EEC on Genetically Modified Organisms', http://europa.eu/rapid/pressReleasesAction.do?reference=IP/96/1148&format=HTML&aged=1&language=EN&guiLanguage=en.

Falkner, R. and Gupta, A. (2009) 'The Limits of Regulatory Convergence: Globalization and GMO Politics in the South', *International Environmental Agreements*, 9 (2): 113–33.

Geuhlstorf, N. P. and Hallstrom, L. K. (2005) 'The role of Culture in Risk Regulations: A Comparative Case Study of Genetically Modified Corn in the United States of America and European Union', *Environmental Science and Policy*, 8: 327–42.

Goffman, E. (1974) *Frame Analysis*. New York, NY: Harper.

Jasanoff, S. (1995) 'Product, Process or Programme: Three Cultures and the Regulation of Biotechnology', in M. Bauer (ed.), *Resistance to New Technology*. Cambridge: Cambridge University Press (311–35).

Lieberman, S. and Gray, T. (2006) 'The So-called "Moratorium" on the Licensing of New Genetically Modified (GM) Products by the European Union 1998–2004: a Study in Ambiguity', *Environmental Politics*, 15 (4): 592–609.

Lieberman, S. and Gray, T. (2007) 'The role of myths in the conflict between the USA and the EU over genetically modified organisms', *European Environment*, 17 (6): 376–86.

Lieberman, S. and Gray, T. (2008a) 'The impact on developing countries of the tension between the USA and the EU over GMOs: opportunity or threat?' *British Journal of Politics and International Relations*, 10 (3): 395–412.

Lieberman, S. and Gray, T. (2008b) 'The World Trade Organisation's Report on the EU's Moratorium on Biotech Products: the wisdom of the US challenge to the EU in the WTO', *Global Environmental Politics*, 8 (1): 33–52.

Majone, G. (1996) 'Regulation and its Modes', in G. Majone (ed.), *Regulating Europe*. London: Routledge (9–27).

Marris, C. (2000) 'Swings and Roundabouts: French Public Policy on Agricultural GMOs 1996–1999', *Cahiers du C3ED*, No 00–02 February 2000.

Meacher, M. (2004) Interview with Michael Meacher – Former Environment Minister, April 2004, Westminster.

Miller, H. (1999) 'Substantial Equivalence: its uses and abuses' *Nature America*: 1042–3, available at http://www.colby.edu/biology/BI402B/Miller%201999.pdf.

Office of Science and Technology Policy (1986) 'Coordinated Framework for Regulation of Biotechnology' *AGENCY: Executive Office of the President, Office of Science and Technology Policy*. June 26 1986 (available at http://usbiotechreg.nbii.gov/Coordinated_Framework_1986_Federal_Register.html, accessed 12 January 2011).

Patterson, L. A. (1997) 'Regulating biotechnology in the European Union: Institutional responses to internal conflict within the Commission', in *European Union Studies Association (EUSA) Biennial Conference, 29 May – 1 June 1997*. Seattle, WA (available at http://aei.pitt.edu/2700/01/002756_1.PDF).

Rein, M. and Schön, D. (1993) 'Reframing policy discourse', in F. Fischer and J. Forester (eds), *The Argumentative Turn in Policy Analysis and Planning*, London: UCL Press Limited and Duke University Press (145–66).

Roy, A. and Joly, P. B. (2000) 'France: broadening precautionary expertise?' *Journal of Risk Research*, 3 (3): 247–54.

Schön, D. and Rein, M. (1994) *Frame Reflection: Towards the Resolution of Intractable Policy Controversies*. New York: BasicBooks.

Strauss, D. (2008) 'Feast or Famine: The Impact of the WTO Decision Favoring the US Biotechnology Industry in the EU Ban of Genetically Modified Foods', *American Business Law Journal*, 45 (4): 775–826.

Thatcher, M. (2002) 'Analysing Regulatory Reform in Europe', *Journal of European Public Policy*, 9 (6): 859–72.

Trachtman, J. (2007) 'Embedding Mutual Recognition at the WTO', *Journal of European Public Policy*, 14 (5): 780–99.

UN (1982) World Charter on Nature, http://www.un.org/documents/ga/res/37/a37r007.htm.

US Biotech Regulation (2010) 'Frequently Asked Questions', *US Regulatory Agencies Unified Biotechnology Website* (http://usbiotechreg.nbii.gov/FAQRecord.asp?qryGUID=3, accessed 12 January 2011).

Vogel, D. (1997) 'Trading up and Governing Across: Transnational Governance and Environmental Protection', *Journal of European Public Policy*, 4 (4): 556–71.

Vogel, D. (2003) 'The Hare and the Tortoise Revisited: The New Politics of Consumer and Environmental Regulation in Europe', *British Journal of Political Science*, 33: 557–80.

Wach, M. (2005) Interview with Michael Wach, USDA APHIS, 7 November 2005, Washington DC.

Wiener, J. B. and Rogers, M. D. (2002) 'Comparing precaution in the United States and Europe', *Journal of Risk Research*, 5 (4): 317–49.

WTO (2006a) 'European Communities – Measures Affecting the Approval and Marketing of Biotech Products: Interim Reports of the Panel', 7 February 2006. Available online at *GeneWatchUK*, (http://www.genewatch.org/sub.shtml?als[cid]=405264, accessed 10 May 2006).

WTO (2006b) 'Understanding the WTO: Settling Disputes: A Unique Contribution', *World Trade Organisation*, (http://www.wto.org/english/thewto_e/whatis_e/tif_e/disp_e.htm, accessed 5 May 2006).

WTO (2006c) 'Understanding the WTO: The Environment – A Specific Concern', *World Trade Organisation*, (http://www.wto.org/english/thewto_e/whatis_e/tif_e/bey2_e.htm, accessed 5 May 2006).

WTO (2006d) 'European Communities – Measures Affecting the Approval and Marketing of Biotech Products – Final Reports of the Panel', *World Trade Organisation*, (http:// docsonline.wto.org/imrd/GEN_searchResult.asp, accessed 29 January 2007).

7 The global battle over the governance of agricultural biotechnology

The roles of Japan, Korea, and China

Yves Tiberghien

Introduction

In the global battle over the governance of agriculture biotechnology (commonly referred to as genetically modified organisms, or GMOs), what has been the position of East Asia? Given the sharp division between the 'substantial equivalence' approach of North America and the 'precautionary principle' advocated by Europe, where did countries such as Japan, Korea, and China end up?

A comparative look at Japan, Korea, China, and even Taiwan, highlights three paradoxes. First, although they all share a strong developmental state, have targeted biotechnology as a priority industry for support, and have a strong agriculture dependence on US exports, all three of them experienced a similar reversal on GMOs. At different speeds, they ended up espousing the principle of precaution, ratifying the Cartagena Biosafety Protocol, and adopting mandatory labelling. Second, if it is clear that the shifts in Japan and Korea are connected to effective non-governmental organisation (NGO) actions that changed public opinion and led to political reactions, why did China follow a similar path, given China's non-democratic regime? Third, the legal moves taken by these three countries do not seem to have deeply affected trade flows, in contrast to the well-known disruptions in North American–European trade. How was this possible?

This chapter builds upon theories of social movements and institutional change to analyse the shifts in East Asian positions. It argues that the Japanese and Korean reversals were caused by tensions between a rising civil society and existing bureaucratic structures, leading to political arbitrage in favour of public opinion. In both cases, however, the regulatory phase remained dominated by traditional actors, who managed to shape legislation such that loopholes and thresholds allowed trade flows to remain unaffected. The chapter further argues that China, surprisingly, followed a similar pattern, thanks to a process of non-institutionalised representation and extensive deliberation within an autocratic regime. Public opinion among urban consumers has converged in the three countries to a large extent. And regulatory outcomes have also converged, largely around an EU inspiration and Japanese practical adaptation.

These findings have several important implications. First, agriculture biotechnology is a multifunctional issue area that touches upon complex intersections between traditional concerns for competitiveness and novel concerns for transparency, accountability, and long term sustainability. Traditional policy making networks and institutions are not well positioned to resolve such tensions and political arbitraging has become necessary in all systems, be they democratic or autocratic. Second, institutionalised democracy is not the only route for participation. In many ways, the degree of public participation and deliberation in policy making concerning GMOs has been higher in China than in Canada or the US, where private interests have dominated the regulatory process. Finally, the battles over governance of first generation GMOs are only a first step. The next battles over cloned beef, GM fish, GM animals, and pharmacrops are likely to be even more divisive and politicised in China, given the absence of currently suitable institutions to reconcile concerns for progress and competitiveness with concerns for long term public goods such as sustainability, health, and democratic accountability.

The rest of this chapter proceeds in three sections. The next section provides a general theoretical framework. The following section briefly reviews the Japanese case. The last section applies a comparative framework to the Chinese case before the chapter closes with conclusions.

Theoretical framework

Social movements, tipping points, and institutional change

This chapter builds upon insights from theories of social movements and on social/norm entrepreneurs (in particular, Ellickson 2001; McAdam, Tarrow and Tilly 2001; Tarrow 1998; Tilly 2004). Key relevant elements coming out of these theories include the concepts of *collective challenge* (Tarrow 1998) and *repertoires* (Tilly in McAdam *et al.* 2001), as well as the importance of *political opportunities* (McAdam, Tarrow and Tilly 2001). Coming from the field of economics, work by Ellickson (2001) on entrepreneurial norms, particularly the work of pioneers in the creation of meanings and new norms, is also very useful.

Applying these concepts to the political economy of East Asian countries, I argue that a regulatory regime shift from a traditional insider-based model to a new fluid model that incorporates bottom-up elements and public good concerns in a field such as GMOs can occur as the outcome of contention between three forces:

- the actions of social entrepreneurs (namely civil society) in issue areas where public opinion (and media) are receptive;
- a political opening, such as party politics in flux (Japan), an increase in the role of urban politics (Japan and Korea), or weakening bureaucratic control (Japan and Korea); and,
- a process of decentralisation with increased competition between local and central governments (Japan).

In such situations, the initial equilibrium of delegation to insider bureaucratic structures based on the triumvirate of bureaucrats, scientists, and business interests becomes politically untenable. A tipping point may occur, leading to political intervention, typically by political first movers (political entrepreneurs) concerned with electoral interests in urban constituencies.

This process, however, occurs within a first phase of agenda setting. It is followed by a phase of regulatory politics in which traditional actors – bureaucrats, interest groups, and scientists (the latter often with links to industry) – remain dominant. It leads to a situation where those traditional actors have received a political mandate to act differently but retain their longstanding interests. The outcome, unlike the European case, is one of broad change in the framework with secondary level technical specificities that do not end up disrupting trade and industrial interests.

Nonetheless, for countries like Japan and Korea, this regulatory shift in the field of GMOs represents a large change and possibly heralds a new model of governance. It represents the disruption of regulatory politics (so-called iron triangle policy making) pursued since at least the 1950s by new, civil society actors.[1] In the case of GMOs in Asia, these new actors tend to be primarily consumer groups acting in alliance with some emerging environmental groups (more so in Korea), occasional religious groups (Korea), minority small-scale or organic farmers (both Korea and Japan), and urban politicians. These interests were traditionally excluded from traditional regulatory politics in both Japan and Korea.

At the same time, so far it is clear that the emerging civil society has only managed to capture the political agenda (taking advantage of more competitive party politics and democratic competition in both cases), and has not shaped the detailed crafting of final regulations. This leads, presently, to an unstable outcome: merely partial and fragile regulations, which lack complete legitimacy and can be seen as a transitional compromise.

Alternative models

One common argument is that responses to GMOs are tied to culture. According to such an argument, European attachment to quality food explains the gut-level rejection of GMOs. Gottweis (1995) and Andrée (2002) have argued that the value framing of science in different settings dominates the political response to GMOs. Cultural arguments are often used in public forums and in the press. However, how can a cultural explanation account for the major and sudden change of direction in GMO policy? Indeed, cultural norms, meanings, and beliefs are developed over long periods of time and are unlikely to change suddenly. They provide foundations for stable culture-specific institutions and are useful to explain cross-national differences. However, they are not good explanations for rapid policy shifts within a given culture.

Scholars and policymakers alike have also suggested that the EU and Japanese actions were driven by trade considerations and showed a return to some degree of protectionism against US imports.[2] In 1999, the new EU trade commissioner,

Pascal Lamy, met with US President Clinton and was later 'grilled' by the US Congress. Both meetings focused partly on the GMO regulations, which were clearly framed as an intentional trade obstacle for domestic purposes (Lamy 2002: 123). However, studying the Japanese GMO regulatory response through a trade lens is clearly flawed. Japan depends on trade to a much larger extent than the US, and has fully supported and led the attempt to further the trade agenda in various WTO meetings. There is no competitive advantage to Japan in restricting imports of GM crops. The only impact is a temporary redirection of trade flows from GM exporters to GM-free crop exporters. In fact, Japanese GMO regulations are having a negative economic impact on Japan, undermining Japanese technological leadership in the field of biotechnology for decades ahead. Clearly, a focus on business and trade interests yields a weak explanation. Business interest groups and their networks with ministries were stacked against GMOs in the Japanese case.

Other arguments have emphasised the impact of one single event: the 'mad cow' (BSE) crisis in the UK and later in the EU (Vogel 2001; Vogel and Cadot 2001). According to such views, the mad cow crisis dealt a terrible blow to the perceived safety of food regulation in Europe and turned Europeans and Japanese against GMOs. However, the correlation between GMOs and BSE does not work well in the Japanese case. The large wave of popular actions and consumer actions in Japan dates back to 1997–8, while its BSE crisis only started in 2001. In addition, Japanese NGO networks have built upon arguments that were independent from the BSE crisis in their fights against GMOs, focusing on issues such as the patenting of life, the threat to biodiversity, and the risks of mutations.

Adaptation to the case of China – representation without institutions

Similarly, in China, it is hard to understand why the balance tilted toward more precaution without taking into account the role of public voice. I argue that the Ministry of Agriculture and the State Council decided to move toward precaution primarily because of a choice to integrate the opinion and interests of urban consumers, as channelled by NGOs, consumer associations, the media, and key academics. This process of integration of grassroots public good concerns from the middle class followed what I call a process of *representation without institutions*. It led to an outcome that is relatively similar to that of the EU and Japan, despite the absence of democratic institutions that were critical in those systems. The outcome, however, may be temporary as the channels for public voice integration are not institutionalised.

At a broad level, this argument is compatible with work done by He and Warren on authoritarian deliberation in China (He and Warren 2011). They argue that new participatory innovations have introduced key deliberative dimensions (village elections, recall votes, deliberative polls, public hearings, citizen suits, some room for NGOs) in Chinese governance, despite the overall authoritarian framework. The motivation for these innovations, however, is legitimacy building and not empowerment. It focuses solely on issues of governance, but can nonetheless be construed as a form of administrative democracy. This leads, according to He and Warren, to the paradox of deliberative institutions under authoritarian rules. In another recent work, Leonard argued that Chinese governance increasingly integrated dimensions of deliberative dictatorship, defined as the integration of public consultation and elite debates in the policy making process, albeit in a controlled and scientific way (See Leonard 2008, particularly the section on deliberative dictatorship (66–81)). Leonard writes, 'public consultations, expert meetings and surveys are becoming a central part of Chinese decision-making' (2008: 67). In a different form, another recent work by Mertha has shown that new coalitions of NGOs, state officials, and policy entrepreneurs could lead to bottom-up policy change, under the important proviso that the regime does not feel threatened in any fundamental way (Mertha 2008).

In the case of GMO governance, I define here the process of representation without institutions as an interactive process where the interests of the state (pursuit of legitimate governance) and those of new coalitions between grassroots activists and policy entrepreneurs (academics) can meet and lead to participatory governance. This process is facilitated or occasionally even enabled by the existence of inter-ministerial gaps (fragmented authoritarianism), which ensures the absence of a clear elite consensus. It primarily means that the government is seeking to respond to bottom-up public good concerns and integrate preferences of urban consumers in its governance outcome, even if the channels for the expression of this public voice are not formally institutionalised. The process emphasises a primary domestic audience for policy outcomes and a degree of dialogue between state and society.

Japan – political reversal and regulatory adaptation

Unlike Europe, Japan has a very low food sufficiency, at approximately forty per cent overall. The situation is much worse for key international commodities such as soy (less than five per cent self-sufficiency), corn (less than five per cent), wheat (less than fifteen per cent), and canola (zero per cent). In this context, Japan has less autonomy than Europe in its actions relative to its largest supplier by far, the United States. In addition, the government (both Ministry of International Trade and Industry (MITI) and the Ministry of Agriculture) has supported agriculture biotechnology since the mid-1980s as one priority sector for Japanese industrial policy. Hundreds of public tests have taken place annually since 1989, and Japan strongly adhered to the US sponsored Organisation for Economic Co-operation and Development (OECD) consensus around the concept of substantial equivalence. As late as January 1997, Prime Minister Hashimoto testified in the Diet that Japan was content with US safety approvals and would not require mandatory labelling or impose any further restrictions on imports of GM soy or corn from the US.

However, in the midst of a charged global context in 1999–2004, Japan ended up tilting toward the regulatory camp. It passed mandatory labelling regulations in 1999 (effective 2001), thus joining the handful of countries that have imposed such high requirements on the industry (together with the EU, Australia, New Zealand, and Russia, and later followed by Korea, China, and others). Japan further raised the bar for the approval of field tests and several prefectures took further steps to prevent these tests. In addition, Japan signed and ratified the Cartagena Protocol on Biosafety and passed one of the most far-reaching sets of domestic regulations in 2004 to implement the Protocol. This was followed by further regulatory moves at the local level. In 2005, the government of Hokkaido, a key growing prefecture for corn, soy, and wheat, passed more stringent regulations on field tests and production (making it all but impossible to grow GMOs in Hokkaido). This highly surprising outcome represents a major and risky path departure for Japan. This posture not only risks limiting Japan's long term competitiveness in biotechnology, it also further entrenches a global contest over biotech governance. Why then did Japan take the risk of such significant change with global impact?

The argument in short

The initial consensus among large industry, scientists, trade interests connected to the US, the elite bureaucracy, and mainstream politicians on support for agriculture biotechnology found itself under attack after 1996 from a surprising new quarter. A new and rather weak NGO coalition managed to present new information and data, raise new questions that interested the public, and gained access to the media (something rarely achieved by NGOs in Japan, given the strong business interests in the media). The civil society leaders involved originated from rather mainstream consumer associations and focused on health safety and issues of transparency and democratic accountability initially. Environmental concerns were added in a later phase.

These groups organised massive petitions through consumer cooperatives (reaching one or two million signatures by the end) and also used their local grass roots member organisations to lobby local governments. Local governments passed resolutions demanding labelling. Initially, the central ministries and politicians ignored this noise, just as they have done in other issue areas such as nuclear energy or dam construction. However, as the issue became more salient with the public, and started influencing voting intentions in urban constituencies, some key mainstream politicians shifted positions and joined non-mainstream politicians. The trigger for change was a committee in the Diet where the majority shifted in favour of labelling on the back of one key mainstream politician shifting his position. The committee demanded a new law on mandatory labelling, forcing the Ministry of Agriculture to begin the drafting process for a new bill.

The regulatory process itself lasted nearly two years and was more transparent than usual, although NGOs were not given seats in the key ministerial deliberative committee in charge of drafting the new legislation. Still, the twenty-one member committee included one representative from a mainstream housewife's association (shufuren) and one consumer representative not associated with the initial NGO actions. The outcome was large in principle, but diluted in practice by key specifications. Unlike Europe, Japan adopted a five per cent threshold for inadvertent presence of GMOs, a threshold negotiated informally by Kohno Taro (key politician of the Liberal Democratic Party (LDP)) with US agriculture officials and initially proposed by the agriculture trade industry in the US. Japan adopted product-based labelling, rather than process-based labelling, thus not requiring the labelling of products such as soybean oil or soy sauce. Lastly, Japan only applied labelling to a restricted list of items. The assessment process is rigorous but purely scientific; it does not give any voice to civil society groups or socio-economic and ethical concerns. As a whole, Japanese regulations are not trade disruptive. The number of GM products approved for imports is higher than in the EU.

The civil society actors

The 'No GMO! Campaign' NGO federation was created only in 1996 and led by Amagasa Keisuke, with Yasuda Setsuko as other key actor. It crafted a tactical alliance with the Seikatsu Club Consumer's Cooperative Union, a cooperative providing alternative family products (such as milk) for twenty-two million households in Japan. The common ground was the demand for mandatory labelling. There was no principled position on GMOs themselves, as the Seikatsu Club wanted to retain the option of cheaper GMO goods for its members. Other groups followed and joined the informal network in future years: Soybeans Trust, Rice Trust, Slow Food Cafes, etc. After 2001, the alliance included new groups, such as environmental NGOs, organic farmers (a key link in the Hokkaido case), and anti-globalisation NGOs, which are weak in Japan.

Agenda capture mechanisms

The NGO federation managed to attract media and public attention to a level that astonished even itself, given its humble origins and resources. They did this through three main mechanisms:

- The NGO coalition used the very large membership of the Seikatsu Club cooperative to gather two million signatures in a massive petition demanding mandatory labelling for GMOs during 1997–8. The petition was used as a large signal that helped the NGO coalition capture the public's attention and put a very large physical pressure on the Ministry of Agriculture, Forestry and Fisheries (MAFF), as the many boxes of petitions literally swamped the ministry.
- The NGO coalition also successfully lobbied local assemblies to pass resolutions demanding GM labelling. In the end, over fifty per cent of local governments (1600 out of 3000) passed such resolutions. The very first local government to move was the Tokyo assembly, but the wild fire spread quickly.
- As well, NGOs organised public activities in the Diet, such as *benkyokai* (study groups) and lobbying of key members of parliament.

The NGO grouping was also able to use EU blueprints and norms. Amagasa, in particular, was always well plugged-in to the EU NGO community and often attended NGO meetings there.

It is also clear that the NGO campaign was able to gain public attention because it dovetailed with a larger crisis of public trust in bureaucratic governance and political leadership in Japan.

The key political link – the Diet sub-committee on consumer issues[3]

Interestingly, the key political moment where political leadership was exercised happened through a Diet committee. The special committee on consumer issues publicly put the GMO issue on its agenda in December 1997 by forming a special sub-committee in charge of the issue of GMO labelling (消費者問題等に関する特別委員会、遺伝子組み換え食品の表示問題等に関する小委員会). The committee on consumer issues was created during the 1973 oil shock; however, it was inactive for decades. The sub-committee on GMO labelling had debates and hearings throughout 1998 and 1999, parallel to the MAFF drafting committee. It issued a final report in 1999 that demanded legislation on GMO labelling.

The key driving forces behind the labelling committee was a coalition of opposition women MPs coming from the Democratic Party of Japan (DPJ), the Komeito, the Japan Communist Party, and the Shaminto. DPJ member, Ishige Eiko, led this informal coalition. On its own, however, this coalition did not form a majority of the committee. The crucial moment came when a rising star in the ruling LDP, Kohno Taro, joined this coalition and tilted the majority in favour of labelling. This represented a rare expression of interests by urban consumers/voters against producer-oriented politics, a tectonic shift in Japanese policy making.

Tellingly, the committee won its battle but it was a Pyrrhic victory. By attracting so much attention, it also attracted the ire of the party leadership in both the LDP and the main opposition DPJ. There was a tacit understanding at the top level in both parties (probably after lobbying by pro-industry and pro-MITI politicians within the party) to shut down the entire committee in 2001. The window for political leadership and for the formation of an unusual alliance within Japanese politics thus closed.

Continuing contention and unstable equilibrium

On the whole, NGOs and the public showed their discontent about the loose regulations adopted by the Japanese government. In July 2002, a government survey showed that eighty-four per cent of the public wanted labelling even below five per cent, and that seventy-six per cent wanted labelling even for processed goods (e.g. soy sauce) (Frid 2002). In March 2007, the No! GMO Campaign, Greenpeace, Toziba (the young people's group calling for a soybean revolution) and others launched a campaign to collect one million signatures demanding stricter labelling. The campaign included a big concert called 'Earth and Peace Festival' in Shiba park. The government ignored it, however.

Meanwhile, action took place at the local level. In February 2005, the prefecture's government for the big northern island, Hokkaido, passed a regulation tightening requirements for field tests and production. It represented a rare step by a local government to push legislation against the centre. The process also included remarkable institutional innovations, such as intense public consultations, the presence of NGOs in the drafting council, and the balancing of all interests by one neutral academic chair of the committee, Professor Matsui. Also unusual was the coming together of a strong coalition, including urban consumer NGOs and farming interests, behind political entrepreneurs and the central government. Such a dynamic is quite similar to what happened in Switzerland or in regions such as Bretagne in France. Visits by Canadian anti-GMO activist Percy Schmeiser to Hokkaido in 2002 and 2005 played an important coalescing role. Hokkaido's actions influenced other prefectures, and the important rice-growing region of Niigata followed on the Hokkaido blueprint.

Japanese NGOs also had a global reach. In 2004, the No! GMO Campaign put together a hand-delivered petition signed by 413 NGOs, companies, and local consumer unions in Japan representing 1.2 million people. It was addressed to the heads of regulatory authorities in Canada and the US, threatening to boycott Canadian and US wheat if the authorities approved GM wheat. This too appeared to have had a significant impact.

Over the period 2004–06, Japanese delegates joined EU conferences and GMO-free region conferences. In October 2007, the No! GMO Campaign, an alliance of eighty consumer groups with farmers groups and individuals, visited Australia with a petition representing 2.9 million consumers to lobby state premiers to extend moratoria on GMO cultivation.

China – representation without institutions[4]

China and India are increasingly seen as the two key pivotal players with the most potential to affect the future global regime. Given the strong divisions and relatively balanced global coalitions led respectively by the US and the EU, China and India will play a key role in tilting the balance one way or the other. Where China goes, global governance will follow, with significant implications for Canada and the US.

China's position on GMOs is still fluid and evolving. At the same time, the outcome so far is relatively surprising. It is clear that China potentially has a higher stake in the success of GMO technology than any other country due to concerns about food security and diminishing agricultural productivity. Yet, to the surprise of many, China's regulatory stance took a strong turn toward the precautionary principle after 2000. Over the past eight years, China has imposed and enforced mandatory labelling with a demanding zero per cent threshold on many selected products. China held back on approving GM rice until late 2009 and, as a result, almost only grows GM cotton at this point. China also took a strong international position by ratifying the Cartagena Protocol for Biosafety in 2005, a protocol that has not been supported or ratified by Canada and the US. This chapter argues that upward pressures from urban consumers and civil society, not protectionist instincts, have played an important role in swaying the Chinese government toward a regulatory position.

The precautionary position taken by China had initial negative trade impacts for Canada in 2000–01 (e.g. canola), given the abrupt introduction of the approval and labelling requirements. Negotiations held jointly with the US did find a pathway that allowed Canadian exporters to adjust, but the level of canola exports never fully recovered. More importantly, it pits China and Canada in two opposite camps on the global governance of GMOs, beginning with the Cartagena Biosafety Protocol. This is a case where China is siding with the world's majority of countries, and with Europe and Japan in particular, in pushing for global regulations on the trade and labelling of GMOs, while Canada and the US do not, preferring to remain within the framework of the WTO.

China's balancing act – between entrepreneurialism and precaution

China's domestic governance of GMOs and its position in the debate over the global governance of biotechnology represent an interesting balancing act. With its huge population and stagnating agricultural production, China is under strong pressure to use agricultural biotechnology in order to boost productivity. In terms of research, testing, and imports, China is already a strong player in the field. At the same time, China's regulatory position is relatively precautionary and closer to that of the EU and Japan.

The structural pressures toward introducing productivity-enhancing agricultural biotechnology are great in China. China is feeding about twenty per cent of the world's population with less than ten per cent of the world's arable land. China is a huge agricultural producer: with approximately thirty per cent of world rice production, twenty-six per cent of cotton production, sixteen per cent of world wheat production, and thirty-two per cent of canola production, China is the leader for many such commodities.

However, China has recently lost its self-sufficient position and become a major importer of grains and soy. After an early boom stimulated by institutional change in the early 1980s, grain and soy production stopped increasing. Since joining the WTO in 2001, China has become a major player in international agricultural commodity trade. Imports shot up from about US$10 billion in 2000 to nearly US$60 billion in 2008. China was a net exporter of soybeans and corn until 1995. Now, China is the world's largest importer of soybeans and has retained only a barely positive position on corn. China remains largely self-sufficient for wheat and rice. However, China anticipates losing its position of self-sufficiency in wheat, rice, and corn by about 2020, unless it drastically increases productivity. Rice is particularly vulnerable, given the strong limits on China's water supply.

In response to these challenges, China has put a strong emphasis on research in agricultural biotechnology, spending about US$120 million annually from

2000–06. In July 2008, the State Council upped the ante by approving a special science and technology fund of Yuan 20 billion (US$2.9 billion) for research on new varieties of GM crops between 2008 and 2010.

China has developed the largest plant biotechnology production capacity outside the US. It has tested GM technology with novel traits in rice, wheat, potatoes, peanuts, and many others that are distinct from research done in all other countries. The figures from the Biosafety Office under Ministry of Agriculture (MOA) show that 2361 experiments on GMOs were approved from 2002–07.

All Chinese research is public. Nearly fifty universities and research institutes, as well as one hundred and fifty local laboratories are involved in the process.

Imports of GMOs by China

China has authorised the import and processing of a large number of GM crops, including five types of cotton traits, two types of soybeans, seven types of canola, and ten types of corn. China has become the largest importer of GM soybeans in the world. In 2007, China imported from the US alone over US$ 4.1 billion in soybeans (twelve million metric tons), representing forty-one per cent of US soy exports (Becker 2008). By 2011–12, these figures are expected to have more than doubled: twenty-seven million metric tons imported from the US, representing forty-five per cent of total Chinese imports of fifty-eight million metric tons (DowJones 2011). Nearly ninety per cent of US soy production is GMOs. China is also a large importer of GM cotton from the US and elsewhere ($1.46 billion of imports from the US in 2007) (Becker 2008), but it is so far an insignificant importer of GM corn.

For Canada, the most relevant GM crop is canola, most of which is GMO in Canada. China was a major importer of canola from Canada in the late 1990s and early 2000s (absorbing 31–9 per cent of Canadian exports), but that share collapsed in the wake of the introduction of new GM regulations in China in 2001 (Canola Council 2010).

Chinese production of GE crops

China has also become a major player in the production of GMOs since 1997, with six crops approved for production so far (cotton, tomatoes, sweet peppers, chilli peppers, petunias, and papayas). In November 2009, China issued biosafety certificates authorising production for two traits of GM rice and one trait of GM corn. However, as of May 2011, the last required step of authorisation for commercial production was being delayed, while further field experiments were going on. In reality, public concern increased massively in 2010 and 2011 (including in the public media and public debates in the National People's Congress) and the government decided to further delay.[5] To date, therefore, only cotton has been widely adopted by farmers, with other crops not being produced or only in extremely small amounts. As of 2007, it is estimated that sixty-nine per cent of all of China's cotton acreage is GMOs (3.8 million hectares out of 5.5 million hectares in total) (Petry and Wu 2008). A rough estimation is that Monsanto and other foreign firms have about sixty per cent global market share in cottonseed, while Chinese producers have the other forty per cent.[6] Chinese farmers have responded very rapidly in adopting new technology.

These figures need to be placed in perspective. The acreage devoted to GMOs in China has been constant for several years, reaching 3.8 million hectares in 2008 and falling to 3.5 million hectares in 2010 (ISAAA 2011). This figure represents less than 2.4 per cent of the total world acreage devoted to GMOs. In 2000, China was the fourth largest GMO producer in the world. However, by 2008 it had slipped to the sixth rank. Even India reached 7.6 million hectares occupied by GM production. By 2010, China remained in the sixth position, but India had jumped to the fourth position with 9.4 million hectares (all cotton) (ISAAA 2011).

GMO governance in China – strong precautionary elements

The production data noted above show there is strong resistance to the rapid adoption of new GM technology in China, despite the strong economic pressures in the other direction. Most importantly, no single major food crop (rice, wheat, corn, and soy) has so far been approved for GM production in China. It was long expected that China would be the first country to authorise GM rice, (Iran did so in 2007). However, the delays are indicative of the political sensitivity of these issues and battles, concerning the authorisation of GMOs, that have prevailed over the last 5–8 years and were postponed again in January 2008. In late November 2009, the Biosafety Office finally issued safety certificates for two strands of GM rice.[7] Yet, this step was not the final one before production could begin. As of May 2011, reports indicate that the ministry had postponed authorising production, due to increased public concerns and even involvement from the top leadership in the standing committee of politiburo.[8]

The MOA's Biosafety Office is the key actor in Chinese GMO governance, particularly when it comes to the granting of safety certificates for importing or producing GMOs. It operates under regulations promulgated in 2001 by the State Council and updated in 2009. However, other ministries are also involved, particularly the Ministry of Environmental Protection (MEP) as the lead ministry on the Cartagena Protocol for Biosafety, the Ministry of Science and Technology (MOST) regarding research funds, and the Ministry of Commerce (MOFCOM) for trade issues. Ultimately, the State Council is the arbiter of key decisions. The 2001 regulations require both food safety and environmental safety tests before granting a safety certificate, a position that puts China closer to the EU than to the US or Canada.

The other key precautionary component of China's approach to GMO governance is the implementation of mandatory labelling after 2001. The initial decision was taken suddenly by the State Council and caused significant trade frictions with both the US and Canada. The imposition of mandatory labelling with a zero per cent threshold puts China in the same camp as the EU and Japan. However, like Japan and unlike the EU, China restricts the applicability of mandatory labelling to specific products: soybean seeds and oil, corn seeds and oil, canola seeds and oil, cotton seeds and tomatoes. Papayas, for example, are excluded. In addition, China has taken a middle position on the scope of labelling by only requiring the labelling of raw seeds or seeds that have undergone primary processing only (e.g. soybean tofu or oil), but not secondary processing (e.g. chocolate bars with soy lecithin or cookies made with soy oil). Such a position is different from the strict process-based labelling in the EU and more similar to Japan's pragmatic labelling approach. It allows China to avoid labelling the bulk of soybean imports from the US once they enter the processing circuit. Preliminary reports from Greenpeace are that labelling guidelines are generally enforced in China.[9]

The last plank in China's evolving precautionary approach was the ratification of the Cartagena Biosafety Protocol (CBP), a process accomplished in April 2005 under the leadership of the then State Environmental Protection Agency (SEPA). This process of ratification has motivated the MEP to seek an overall Biosafety Law, although it has not yet happened. The ratification of the CBP was a major move that indicated support for a global regulatory approach and for the precautionary approach in general. This decision located China closer to the EU and Japan in terms of principles than to the approach pursued by the US and Canada. It is also important to note that China was faster in ratifying the CBP than its neighbour Korea, despite the strong civil society and democratic pressures to that intent. Korea finally ratified the CBP in 2008.

What lies behind the split in China's approach to GMOs and its shift toward a more precautionary approach in 2001 and 2005?

The key element in this process is public opinion. Interestingly, public opinion is divided. It has been widely reported that Chinese opinion as a whole is both ill informed about GMOs and open to them. For example, the USDA's 2008 biotechnology report contends that 'China's consumers are by and large open to and accept biotechnology products' (Petry and Wu 2008). This report cites a study showing that sixty per cent of consumers nationwide were willing to purchase biotech foods without any price discrimination, while seventy-five per cent of Chinese respondents as a whole were not much aware of GM food.

Greenpeace's own survey of urban consumers in three large cities (Beijing, Shanghai, and Guangzhou) paints a very different picture, even though it is challenged by pro-GMO scientists on the count of sample problems.[10] It found that sixty-five per cent of the consumers show a preference for non-GM in all four main categories surveyed (soy oil, rice, other food products of a plant origin, and food products from an animal origin). Specifically, seventy-nine per cent choose non-GM soy oil over GM soy oil; seventy-seven per cent choose non-GM rice over GM rice; eighty-five per cent choose non-GM food from an animal origin over their GM counterparts; and eighty per cent of the consumers oppose the use of GM ingredients in baby food.[11] If confirmed over other polls, these results could indicate a growing convergence of Chinese urban consumers and middle class with urban consumers in Korea or Japan, itself a very interesting trend.

The MOA's Biosafety Office confirms both results, indicating that the government is aware that there are widely diverse opinions in China. But it is particularly aware of 'rising public awareness', particularly in large cities. The MOA conducts its own investigations and surveys. It also organises its own workshops and study groups. It is aware of Greenpeace's results and confirms that they have reproduced them. The office confirms that labelling came to be seen as 'the right to know for consumers'.[12]

Channels of public voice in Chinese GMO governance

Interviews with various actors point to four channels for the integration of the public voice of urban consumers and the middle class into GMO governance:

1   Space for NGOs (Greenpeace): Greenpeace has been able to operate in Beijing (and Hong Kong) with relative effectiveness. It understands that it can only have a role if it keeps within key boundaries and does not challenge the government head-on. It has focused its work on collecting data and working with academics to produce quality reports (consumer surveys, analysis of patent issues, and so on). Yet these reports are widely read both by academics and by the MOA and MEP. It seems that the government accepts Greenpeace's role of information provision and input into governance. Greenpeace adds that it is relatively at ease with government regulations, given that the government has moved toward precaution and seems to be implementing labelling regulations quite carefully.

2   Consumer Council: This government controlled broad-based organisation gathers information from consumers at the grassroots level and passes it on to higher levels and to the central government. They also organise training on risk management and information sessions for the public. Informally, interviews have indicated that council members are aware of misgivings among urban consumers and have passed on that information to central authorities.

3   Academic Entrepreneurs: Several academic entrepreneurs play a key role interfacing with the international environment and as vectors for representation of public interests. Academics play a key role on the other side of the debate by linking with industry and economic interests and representing those interests to the government.

4   International Norms (from EU and Japan): An interesting input in governance, even if it is an international one, is the influence of international norms. They filter down to some actors and NGOs and come back up to the government. My interviews at the MOA, MEP, and with key academics involved in the two ministries confirm that the Chinese government has thoroughly studied regulations from both the EU and Japan before writing its own rules on safety assessment and labelling. The Cartagena negotiations also provided ample occasions for exchanges and socialisation.

A delicate balance between two multi-level coalitions

In sum, GMO governance in China involves an interesting confrontation between two large multi-level coalitions. With GMO governance, we are facing a fluid and fragmented governance situation, rather than that of a unified mercantilist state for example.

Conclusion

This chapter has highlighted two puzzling dimensions of GMO regulations in north-east Asian countries. First, China, Japan, and Korea have customarily been seen as traditional developmental states with a strong bureaucratic interest in fostering industrial interests, particularly in new technologies with high potential for competitiveness. However, all three have taken a turn toward a tight regulatory framework and precaution in the case of GMOs. All three have implemented demanding approval mechanisms that include both health safety and environmental impact assessments. All three have adopted mandatory labelling (unlike the US or Canada). All three have ratified the Cartagena Protocol (Japan in 2002, China in 2005, and Korea in 2008). And, all three have had significant trade frictions over GMOs with the US, particularly in the regime creation phase (1999–2002).

This chapter has argued that the origin of this regulatory shift toward precaution and labelling was not the result of traditional interest group lobbying, protectionism, ministerial politics, or even culture. Rather, in both Japan and Korea, unprecedented civil society coalitions came together, capturing the agenda-setting role and forcing the bureaucracy to initiate a regulatory shift. In Japan, the trigger came through local governments, shifting positions by urban politicians (after lobbying by civil society or consumer groups, and the decision of a key committee in the Diet). In Korea, the trigger was a direct link between NGOs and the Minister of agriculture under President Kim Dae-Jung, which provided a particularly propitious political window in 2000–01. Thus, in both Japan and Korea, the change took place through a linkage between political entrepreneurs and civil society. It was the result of political interventions in the regulatory realm. However, in both Japan and Korea, bureaucratic and industrial interests managed to dominate the subsequent regulatory phase and shape the final regime, with resulting loopholes and fine print ensuring minimum disruption of trade and economy.

Second, this chapter has argued that a relatively similar process took place in China despite the absence of democratic institutions. In China, the State Council decided to balance traditional industrial and developmental interests with a growing concern for good governance and legitimate outcomes. Strong demands for mandatory labelling and a tight regulatory regime among the Chinese urban middle class were made clear through opinion polls, consumer information, and NGO activities. The government thus realised that it had to take a middle path. I have called this a process of 'representation without institutions'.

These three cases highlight how GMO regulations are multifunctional issues that have the potential to mobilise unusual grassroots coalitions and public interest even in countries where developmental interests have traditionally dominated. Much more such mobilisation can be expected regarding the next frontier of biotechnology, including GM animals and pharmacrops. Regulation in such fields will be increasingly politicised and publicised and will thus not follow traditional regulatory processes. In all three Asian countries considered here, new institutions are probably necessary to reconcile developmental and public good interests in an efficient and legitimate way.

Notes

1 This approach fits broadly within an emerging new literature on civil society in Japan (Haddad 2007; Hasegawa 2004; Pekkanen 2006; Schwartz and Pharr 2003).
2 Cf. presentation by Dr. Chris Sommerville, Director of the Carnegie Institute and Professor of Biology at Stanford University, on 15 February 2001 at Stanford University. See also the international trade lens used in Falkner 2000 and Newell and Mackenzie 2000.
3 This is based on personal interviews with key members of the committee and government officials, including interviews with members Kohno Taro and Ishige Eiko in 2005 and 2006.
4 This section draws upon a previously published working paper Tiberghien, Yves (2010) *The Global Governance of Biotechnology: Mediating Chinese and Canadian Interests*. CIC (Canada International Council) China Papers. No. 13. July.
5 Source: personal interview with informed scientist, Beijing, May 2011.
6 Source: personal interview with informed scientist, Beijing, May 2008.
7 Source: Personal interview with informed NGO leader, Beijing, December 2009.
8 Anonymous Source: Personal interview in May 2011.
9 Personal interview, Greenpeace, May 2008.
10 Personal interview with Huang Jikun, 15 May 2008.
11 Survey results shown during interview at Greenpeace's office, Beijing, 15 May 2008.
12 Personal interview with head of Biosafety unit, Beijing, 15 May 2008.

Bibliography

Andrée, P. (2002) 'The Biopolitics of genetically-modified organisms in Canada', *Journal of Canadian Studies* 37(3):162–91.

Becker, Geoffrey (2008) *CRS Report for Congress: Food and Agricultural Imports from China*. Washington DC: Congressional Research Service.

Canola Council, 'Seed Exports (Historic)' (1 August 2000 – 31 July 2010), http://www.canolacouncil.org/seed_exports_historic.aspx (accessed June 2011).

DowJones News Wire, 9 March 2011, 'China to import record soy totals, USDA says', http://www.agriculture.com/markets/analysis/soybeans/china-to-impt-recd-soy-totals-usda-says_10-ar15230 (accessed June 2011).

Ellickson, Robert C. (2001) 'The Market for Social Norms', *American Law and Economics Review* 3 (1): 1–49.

Falkner, Robert (2000) 'Regulating Biotech Trade: The Cartagena Protocol On Biosafety', *International Affairs* 76 (2): 299–313.

Frid, Akiko. 8 July 2002, *Gentech Archives*, 'Japan-New opinion poll: 80 per cent unsatisfied with the current GMO labeling', http://www.gene.ch/gentech/2002/Jul/msg00075.html (accessed June 2011).

Gottweis, H. (1995) 'German politics of genetic engineering and its deconstrucrion', *Social Studies of Science*, 25(2): 195–235.

Haddad, Mary Alice (2007) *Politics and Volunteering in Japan: A Global Perspective.* Cambridge, UK; New York: Cambridge University Press.

Hasegawa, Koichi (2004) *Constructing Civil Society in Japan: Voices of Environmental Movements*. Melbourne, Vic.: Trans Pacific.

He, Baogang and Mark E. Warren (2011) 'Authoritarian Deliberation: The Deliberative Turn' in *Chinese Political Development Perspectives on Politics* 9 (2): 269–89.

International Service for the Acquisition of Agri-Biotech Acquisitions (ISAAA) (2011) *Annual Briefs for 2008 and 2010*, http://www.isaaa.org/resources/publications/briefs/42/executivesummary/default.asp (accessed June 2011).

Lamy, Pascal (2002) *L'Europe en premiere ligne*. Paris: Editions du Seuil.

Leonard, Mark (2008) *What Does China Think?* New York: Public Affairs.

McAdam, Doug, Sidney G. Tarrow and Charles Tilly (2001) *Dynamics of Contention*. New York: Cambridge University Press.

Mertha, Andrew (2008) *China's Water Warriors: Citizen Action and Policy Change*. Ithaca, NY: Cornell University Press.

Newell, Peter and Ruth MacKenzie (2000) 'The Cartagena Protocol on Biosafety: Legal and Political Dimensions', *Global Environmental Change* 10: 313–7.

Pekkanen, Robert (2006) 'Japan's Dual Civil Society', *Members Without Advocates*. San Francisco: Stanford University Press.

Petry, Mark and Bugang Wu (2008) 'Usda Foreign Agricultural Service – Gain Report' *(Global Agriculture Information Network)*. GAIN Report. Beijing.

Schwartz, Frank J. and Susan J. Pharr (2003) *The State of Civil Society in Japan*. Cambridge, UK ; New York, NY: Cambridge University Press.

Tarrow, Sidney G. (1998) *Power in Movement: Social Movements and Contentious Politics*. Cambridge, UK; New York: Cambridge University Press.

Tilly, Charles (2004) *Social Movements, 1768–2004*. Boulder: Paradigm Publishers.

Vogel, David (2001) 'The Regulation of GMOs in Europe and the United States: A Case-Study of Contemporary European Regulatory Politics', *Publication of the Study Group on Trade, Science and Genetically Modified Foods*. Washington, DC: Council on Foreign Relations.

Vogel, David and Olivier Cadot (2001) *France, the United States, and the Biotechnology Debate*. Washington, DC: The Brookings Institution.

8 The EU's governance of plant biotechnology risk regulation

Still contested, still distinct

Paulette Kurzer and Grace Skogstad

Currently, European Union (EU) standards concerning the release of genetically modified organisms (GMOs) for plant cultivation and for consumption as food or animal feed are the strictest in the world, and differ in several important ways from those in the United States, the world's leading producer of genetically modified (GM) crops. Starting from the principle that GMOs are novel entities, the EU requires every GM product to undergo a scientific risk assessment to demonstrate its safety before it can be licenced for sale. It also requires all seeds, foods, and animal feed derived from GMOs to be labelled. Moreover, traceability rules require all those who produce, process, or market a GMO in the EU to keep records to enable the GMO to be tracked as it moves through the food and feed supply chain. None of these requirements exist in the US. The transatlantic EU–US gap resulting from the more rigorous EU GMO standards has also created a large EU–US gap in commercialisation of GM crops. While the US dominates in global GM crop production (growing forty-five per cent of GM crops worldwide – of which corn, soybeans and cotton are the USA's largest crops), the EU's twenty-seven member states grow less than one per cent of GM crops grown worldwide (ISAAA 2011).

The EU's GMO regulatory framework has created tensions with foreign countries that produce and export GM crops and feed – the United States, Argentina, and Canada – and who find it difficult, if not impossible, to export their GM products to the EU. These external trade pressures persist, as the EU has failed to bring itself into full conformity with a 2006 World Trade Organization (WTO) panel ruling that EU member states' measures to keep GMOs out of their countries were inconsistent with WTO law (Bernauer and Aerni 2007; Falkner 2007; Levidow 2007; Ramjoué 2007; Pollack and Shaffer 2009; Winickoff *et al.* 2005; Young 2009). Also persisting is the longstanding contestation inside the EU around GMO risk regulation (Bernauer and Caduff 2004; Ferretti and Pavone 2009). Amidst continuing internal and external controversy about GMO risk regulation procedures, an important and timely question is whether this controversy and pressures for change will result in EU regulatory changes that converge on the more permissive American standards and close the regulatory gap with the US.

We address this question here. To do so, we begin in the next section with an account of the EU's stringent regulatory approach to plant biotechnology (GM crops and foods). We describe the foremost features of current EU GMO risk regulation and trace them to, first, an EU political culture that is overwhelmingly sceptical of the benefits of GM foods and worried about their safety; and, second, an institutional framework in which decision making on GMO risk regulation has required consensus building across multiple political institutions and actors whose views on GMO risk regulation differ sharply. These political actors are member state actors in EU level institutions: the Council of Ministers, the different directorates in the European Commission, and the European Parliament. Actors in these institutions have been divided on the merits of GMOs and how best to regulate their risks to the environment as well as human and animal health. They have also struggled to reconcile the differing perspectives of non-state actors representing the biotechnology industry, environmentalists, consumers, and the agricultural and food/feed sectors. The GMO risk regulation framework in place since 2004 responded to the most risk-averse member states and Commission Directorates General (DGs) but it was still a fragile – and ultimately inadequate – compromise among the disparate views of this pluralistic community.

In the following section, we document the continuing controversy as political authorities attempt to licence GM products under the 2004 risk regulation framework, the ineffectiveness of GMO risk management procedures, and the contested legitimacy of EU risk assessments of GMOs by the European Food Safety Agency (EFSA). In the last section, we turn to two developments that have implications for shifting the EU GMO risk regulation framework closer to the more permissive US approach. These are, first, regulations regarding the import of GM products for animal feed that contain small amounts of GM products not yet approved by EU regulators; and, second, the right of individual member states to restrict the cultivation of GM crops in their territory. The effort to resolve controversy surrounding the EU's zero tolerance policy for the import of GM feeds can be read as a modest attempt to reconcile EU regulatory standards with those of GM animal feed exporting countries, including the US. The proposal, which is still under discussion in spring 2011, to overcome the stalemate over the approval of GM seeds for cultivation could close – albeit to a limited degree – the American–EU regulatory gap on approvals of GM crops for cultivation. This proposal devolves greater decision-making authority to individual member states to determine whether to allow GM seeds to be grown in their country. If the proposal is implemented, it is reasonable to expect some member states to permit new GM crops to be grown in their territory. It is also reasonable to expect other member states to continue to refuse to allow the entry or production of GM crops in their country. The consequences of the latter are that the EU will continue to lag behind the US in authorisation of GM crops.

We conclude that none of the EU initiatives to manage internal EU controversy over plant biotechnology will see the EU converging very soon on the American permissive approach to GMOs. Rather, the considerable controversy around GMOs inside the EU, the institutional barriers to change, and the European public's aversion to GM foods, mean that the EU will continue on its distinct GMO risk regulation trajectory.

The EU GMO risk regulation framework

The current EU regulatory framework for appraising the risks of GMOs and licensing GM products for sale has been in effect since 2004. Consisting of several pieces of legislation,[1] it incorporates principles adopted as early as 1990 in the first EU biotechnology directive (Directive 90/220), but also entails new principles and procedures in an effort to overcome criticisms of earlier legislation. Among these criticisms, voiced particularly by environmental groups, was that GMOs were being approved without sufficient scientific attention to their environmental and health risks,[2] and without monitoring of their risks once they were approved and commercialised. Consumer groups criticised the absence of comprehensive labelling of GM products, a regulatory gap that limited the right of consumers to avoid GM foods if they had safety concerns. These and other criticisms, and public opposition to GM products, led a handful of member states to refuse to adhere to EU decisions authorising GMOs throughout the EU. Beginning in 1997, Austria, Italy, and Luxembourg invoked the legislative safeguard clause to ban the import and use within their borders of a GM maize (corn) that the European Commission had approved. Other attempts by the Commission to approve GM products against the wishes of a majority of member states brought the entire GMO regulatory framework to a halt in June 1999 when member states in the Council of Ministers signalled that they would not authorise any new GMOs until new regulatory procedures were in place (Ansell and Vogel 2006; Tiberghien 2009).

The regulatory framework that responded to these concerns established the following procedures for allowing GM products to be sold and/or cultivated in the EU market. First, all GMOs – whether used for cultivation, in animal feed, or for consumption as human food – must undergo a scientific risk assessment to determine their safety to the environment and to animal and/or human health. The task of scientific risk assessment is given to the EFSA; the members of its scientific committees are independent of governments and the European Commission. EFSA uses its scientific risk assessment to provide advice to the European Commission on whether to licence a GMO. While the Commission does not have to heed EFSA's opinion, it must provide an explanation when it fails to heed EFSA's advice (Buonanno 2006).

Second, GM products are approved for licensing and marketing for a specific purpose (such as food, animal feed, or cultivation) in the EU's twenty-seven member countries by representatives of their governments, who act in response to a recommendation from the Commission, (which, as just noted, acts on a recommendation to that effect from EFSA). Since 2010, it is the Directorate-General for Health and Consumers (DG SANCO) in the European Commission that is responsible for initiating all proposals to representatives of member states to approve a GMO variety. It first seeks the 'opinion' (support) of the Standing Committee on the Food Chain and Animal Health (SCoFCAH), on which sit those member state officials with expertise on the subject. If this committee, chaired by a Commission official, supports the Commission's opinion by a qualified majority vote (seventy-five per cent of the votes, or 258 out of 345 votes), the Commission can adopt its draft measure and member states are required to implement it. If the Standing Committee does not approve the product, by producing a blocking minority (ninety-one votes), the Commission's proposal to authorise a GM variety goes to a second institution: the Council of Ministers comprised of elected politicians from member states at the cabinet level. Approvals for GM food and feed are normally considered in the Council that consists of member state ministers responsible for agriculture (the EU Farm Council), while those involving the planting of GM seeds are discussed by ministers responsible for the environment (the EU Environment Council). A qualified majority of the Council is needed to approve or disapprove the Commission's recommendation to licence the GM product. Should the EU Council not muster a qualified majority vote either in favour or against the Commission recommendation, the proposal authorising the GMO 'shall be adopted by the Commission' (Decision 1999/468/EC, Article 5).

Third, member states are individually responsible for the regulation of field trials of GM crops in their country and determine the rules for coexistence of GM and non-GM crops within their borders. They also have 'safeguard' scope to restrict the marketing of an EU authorised GMO within their borders providing that they can furnish 'new or additional information' relating to the human health and/or environmental risk of the GMO since it was given EU licensing consent.

Fourth, all GM food, feed, and seeds are subject to labelling and traceability rules. Labelling allows GM-wary consumers to avoid GM products, while traceability provisions, which necessitate record keeping of all transactions in the movement of GM products through the food/feed chain, allow a GM product to be tracked and recalled in the event of a safety issue (Hu, Veeman and Adamowicz 2005; Noussair, Robin and Ruffieux 2002).

These regulatory standards and procedures were the outcome of the intensive process of consensus building in the late 1990s and early 2000s across EU institutions and political actors. As noted above, reforms were needed to end the suspension of GMO authorisations. The imperative for reforms became more urgent when the US, Canada, and Argentina – all major producers and exporters of GM crops – launched a legal challenge at the WTO, claiming the EU's suspension of GMO authorisations was inconsistent with the EU's WTO obligations. Bridging divisions that were then manifest (and still largely in existence) across the European Commission, member states, and non-governmental organisations, proved to be a lengthy and challenging exercise (Borrás 2007; Skogstad 2003).

Within the European Commission, a schism is apparent between the DGs that are responsible for trade, enterprise/industry, and agriculture, and those responsible for consumer protection and health (SANCO) and the environment. Whereas the former have been in favour of agricultural biotechnology (Young 2010: footnote 15), the latter are more sceptical, if not hostile. DG Trade, which is sensitive to the trade implications of EU restrictions on imports of GM products and the need for the EU to uphold its WTO treaty obligations, has consistently urged more rapid approvals of GM crops. DGs responsible for enterprise/industry and agriculture have been responsive to the concerns of industry groups who argue that restrictive GMO standards impede the adoption of a technology that is needed to enhance their international economic competitiveness. DG SANCO and DG Environment, responsible for initiating GMO legislation and authorisation proposals, have been more responsive to the scepticism and hostility that consumer and environmental groups have expressed about plant biotechnology. The European Commission as a whole is supportive of biotechnology by providing funding for biotechnology research and development, and identifying biotechnology as an element in the Lisbon Agenda to make the EU the most competitive bloc in the world. The Commission has sought to engender the necessary public support to promote biotechnology through GMO risk regulation rules that balance precaution and competitiveness (Pollack and Shaffer 2009; Tiberghien 2009: 395).

Member states are also divided on how best to regulate GMOs. Their differences reflect their distinct political cultures of risk regulation, as well as the structures of their agricultural sectors, the salience of plant biotechnology issues to their public, and the strength of non-governmental organisations and coalitions who support or oppose GMOs (Kurzer and Cooper 2007; Young 2010).

Political cultures of risk regulation, as defined by Jasanoff (2005: 249), incorporate beliefs about the appropriate processes for producing and validating 'credible knowledge claims' about the risks and benefits of plant biotechnology. These knowledge claims are discursively constructed by social and economic actors, reflecting socially embedded values, historic contingencies, and the bias of institutional settings – allowing some, but not other, beliefs about GMO risk regulation to become ascendant (Everson and Vos 2009; Gottweis 1998; Gaskell *et al.* 2001; Toke 2004). The political culture of GMO risk regulation in Sweden, the United Kingdom, and the Netherlands, for example, grants ample authority to scientific methods and scientific knowledge to ascertain the risks of a novel technology like genetic engineering. In contrast to such a scientific rationality political culture (Isaac 2002), Germany is characterised by a precautionary political culture of risk regulation that reflects Germany's history of Nazism, under which scientific ideas were distorted for ulterior political reasons. Such a precautionary political culture – which also characterises Austria – stresses the limits of scientific knowledge and views scientific claims as indeterminate and uncertain, and not necessarily a reliable basis for decisions on how to manage a technology's risks (Borrás 2006; Seifert 2010).

Cleavages in member state approaches to GMOs are revealed in the fact that some countries have banned GMOs in their country while others are allowing GM crops to be grown. Austria (whose citizens voted in a nation-wide referendum in the mid-1990s to ban GMO crops) has the most extensive and comprehensive ban with four varieties of GM rapeseed and one GM corn prohibited (Kurzer 2011). Other countries that currently have bans on EU licenced GM products are France, Germany, Greece, Hungary, Luxembourg, Italy, Poland, and Romania.[3] Countries growing GM crops today are Spain, Portugal, the Czech Republic, Romania, Poland, and Slovakia.[4] Some countries have shifted their positions on

GMOs over time. For example, France initially supported GM crops but banned an EU licenced GM maize (Monsanto's MON810) in early 2008 (Bonneuil, Joly and Marris 2008; Heller 2002).

For many EU member states, the appropriate GMO regulatory policy usually entails difficult tradeoffs because mobilised parts of their publics have different views of plant biotechnology. While public opinion data show that a majority of the public across EU countries does not support GM foods, there are some national differences (European Commission Directorate-General for Research 2010). The Danish and UK publics, for example, are more likely to support GM foods than are their Austrian counterparts. While consumers (with some national differences) are averse to GM foods, farmers, especially those with large operations, are more inclined to support GM crops. Moreover, the governments of countries whose farmers rely on imported feed – as do livestock, hog and poultry farmers in Ireland, the UK, Netherlands, Spain, and Portugal – face pressures to support regulations that ease the import of feed produced from GM maize or soybeans. In a few countries, the unity of perspective across food consumers and food producers eases their governments' task of devising a coherent GMO policy. Austria is a case in point. The anti-GMO perspective of its farm organisations (representing smaller farmers who produce high value-added foodstuffs, including organic food) coincides with that of environmental groups and consumers (Kurzer and Cooper 2007; Tiberghien 2009).

Not surprisingly, given the disparity of views on the risks and desirability of promoting agricultural biotechnology across member states, within the European Commission and among non-governmental actors the GMO regulatory framework, that the EU put in place by 2004, did not satisfy all parties. For those parties, principally the biotechnology industry, that sees genetic engineering as an instrument of competitiveness, the 2004 regulatory burden was too restrictive and would stymie the future of a promising technology. For others, such as the anti-GMO member states who advocated a comprehensive ban on all GM-derived products as well as environmentalists, it was not sufficiently restrictive of GMOs. The framework – characterised as 'the toughest laws on GMOs in the whole world' by David Bowe, member of the European Parliament (cited in Pollack and Shaffer 2009: 343) – acceded to the demands of the more risk-averse and higher standard member states (Vogel 1995). Doing so was viewed to be necessary by the Commission in order to end the de facto moratorium on GMO licensing, restore a functioning internal common market in GMO products, and fend off a US (WTO trade) challenge (Pollack and Shaffer 2009).

The hope that a rigorous GMO regulatory framework – one based on independent scientific risk assessments, labelling and tracking of GM products, and member states' collective right to approve GM products – would end contestation around GMOs has failed to materialise. The implementation of GMO rules to authorise the commercialisation of GM products has been controversial and resulted in few GM products being authorised for sale in the EU. In the next part of this chapter, we document the controversy that surrounds the cultivation of GM crops and the scientific risk assessments of EFSA.

EU regulatory debates and continuing contestation

Since 2004, EU GMO risk regulation procedures have produced at least three noteworthy outcomes. First, only a single new GM variety (a potato) has been approved for cultivation throughout the EU. Shortly after this GM potato was approved, two member states (Hungary and Luxembourg) banned it in their country. This virtual stalemate in the approval of GM plants for cultivation is the consequence of national representatives repeatedly failing to reach a qualified majority in favour of authorising a GM crop for planting in the EU. For example, when SCoFCAH considered maize varieties 1507 and Bt11 on 25 February 2009, only six member states (ninety-one votes) voted in favour of authorisation; twelve (127 votes) voted against; seven (ninety-five votes) abstained; and two (thirty-two votes) were absent (euobserver.com 26 February 2009). In July 2010, the Commission issued import approvals for six GM maize lines. Again, as generally occurs, proceeding ballots in the Standing Committee and in the EU Council of Ministers did not result in the required qualified majority. Since all six GM maize lines (Pioneer 1507x59122, 59122x1507xNK603; Monsanto MON88017xMON810, MON89034xNK603 and Syngenta Bt11xGA21, Bt11) fulfil the legal requirements valid in the EU, the Commission approved their import (GMO Compass 2010). In spite of the repeated approvals of new GM seeds, currently the only GM crop cultivated in any amount in the EU is a variety of GM maize produced by Monsanto – MON810. It was approved in 1998 under the earlier regulatory framework (EU Directive 90/220).

Second, in contrast to the virtual freeze on GM crops, several GMOs have been approved under existing regulatory procedures for import into the EU for use as *animal feed* and/or *processing*. In no case, however, has a GM feed import been approved by a qualified majority of member states. Although it has been unwilling to force GM crops for cultivation on member states, the Commission (acting as the College of Commissioners) has done what EU law requires it to do and has been authorising GM animal feed and/or processing products.

Third, one or more GM products authorised under the existing GM framework – and hence for entry onto the market of each member state – have been banned by some member states. As noted above, these states include large pro-EU countries like France and Germany.[5] The failure of member states to provide the required evidence to justify their bans – that is, new evidence not considered by EFSA regarding the GM variety's health and/or environmental risks – puts these states in contravention of EU law.

Repeated efforts by the Commission to have member states overturn their bans have usually proved futile. So have its efforts to have member states censure their fellow members' bans. In June 2005, when the Commission urged the EU Council of Environment ministers to overturn eight different national bans on GM crop varieties, the Council refused to do so. It did the same thing in 2006 and 2007 when the Commission tried to lift the ban on GM corn in Hungary and Austria. In October 2007, the Council did agree to the Commission's reprimand to Austria for its ban on the use of two types of GM corn for sale as food and feed, while ignoring its ban on GM crop cultivation. However, the vote to ask Austria to lift the ban was split: four member states supported the Commission, fifteen opposed it, and eight member states abstained. In May 2008, the Commission told Austria to lift its ban, and Austria complied a month later. An effort by the Commission to have member states reprimand Austria as well as Hungary for their continuing bans on GM crops in March 2009 failed when an overwhelming majority of member states voted against the recommendation of the Commission and in favour of national discretionary decision making (EurActiv 2009a).

The European Commission has turned to the European Court of Justice (ECJ) to rule that some of these bans are illegal. In March 2011, the ECJ ruled that France's decision to ban the cultivation of the GM maize MON810 was illegal. That same month, the Commission announced it was asking the ECJ to rule on the legality of Poland's prohibition on the production, use, and marketing of a particular GM feed.

Thus far, only one other product has been approved for cultivation – the GM potato for industrial use – although its cultivation has been blocked by national governments. A powerful minority of member states ban GMOs approved at the EU level in their own country while a majority of member states are complicit in their fellow states' illegal behaviour by voting in overwhelming majorities against requiring a member state to lift its ban(s) on the cultivation of EU approved GM crops. This stalemate is indicative of the significant limits to the effectiveness and perceived legitimacy of these same risk management procedures. The latter require oversized majorities of member states to approve or reject a GM variety. The diversity of views across member states and within them has impeded national governments acting coherently on the issue, as they struggle to balance the different interests and competing demands of the biotech industry, consumers, farmers, environmentalists, and other stakeholders. Environmental NGOs have maintained a sustained critique of GMOs' environmental risks, and the public is largely averse to GM foods.

A 2010 report concluded that 'GM food is still the Achilles' heel of biotechnology' in the EU (European Commission Directorate-General for Research 2010: 7). EU citizens' support for GM food, always low, declined over the 1996–2010 period to the point that opponents of GM food outnumber supporters by three to one. While public opinion is more favourable in some member states (Denmark and the United Kingdom), in no EU country does a majority support GM food (Ibid.). These negative public views are rooted in concerns about the safety of GM food, as well as the belief that it has no benefits and is unnatural (Bauer and Gaskell 2002; Meins 2003).

In such a political climate, and in the ever-present competition for legitimacy between the member states and the Commission (Tiberghien 2009), the very authority of the European Commission to enforce the authorisation procedures has, not surprisingly, been challenged; by both member states and anti-GMO groups. Also under attack have been EFSA's risk assessments, on which the Commission relies to affirm the safety of a GMO variety and to determine whether a member state has provided new evidence of a GMO's risks that would support a ban. One criticism is that EFSA's risk assessments have not taken sufficient account of the long-term effects of GMOs, the impact of certain GM corn varieties on non-target species, or national expertise.

In response to criticisms of EFSA's flawed methodologies, the Commission has taken steps to require EFSA to be more accountable to the Commission, member states, and public critics for the quality of its scientific risk assessments. Environment Commissioner Dimas was especially sensitive to charges of failing to take a precautionary approach. On more than one occasion, Dimas asked EFSA to redo its risk assessment of a GM product (AgraFocus 2007; EFSA 2008). In spite of a positive EFSA opinion, DG Environment rejected the authorisation of two maize varieties, 1507 and Bt11, in 2007 and 2008 (EuropaBio 2007). The Environment Commissioner's actions were not, however, backed by the Commission as a whole (ESA, 2008) which affirmed its confidence in EFSA but agreed to refer the two Bt maize varieties back to EFSA for further consideration (European Commission 2008a).

Under member state criticism and pressure, in 2006 the Commission directed EFSA and its GMO Panel to work more closely with member states' risk assessment authorities and to be more transparent in the basis for its opinions. EFSA's GMO Panel was required to take into consideration all the scientific comments from member states prior to finalising its scientific opinion, to explain in more detail when its opinion differed from member state agencies, and to look into the longer-term studies of the environmental and health effects of GMOs (AgraFocus, 2006: 36). In September 2008, member states reiterated the need for 'better cooperation' between themselves and EFSA (AgraFocus, September 2008: 24). In March 2010, EFSA launched public consultations on its environmental risk assessment guidelines and on its draft scientific opinion on assessment of the potential impacts of GM plants on non-target organisms. It followed these consultations with discussions with member state experts. The resulting detailed guidelines, including assessments of the potential effects of GM plants on non-target organisms, may lead to proposals for a regulation (AgraFocus December 2010a: 39).

Still, it is doubtful that these efforts towards more transparent and procedurally accountable risk assessments will satisfy environmental critics. According to environmental organisations such as Greenpeace and Friends of the Earth, EFSA's risk assessments will be flawed as long as the agency fails to admit the uncertainty surrounding the environmental safety of GMOs and adheres to its unwarranted scientific rationality perspective (Friends of the Earth Europe 2004; Levidow 2006).

The controversies that swirl around GMOs and handicap the effectiveness and legitimacy of both EU level risk assessment and risk management of plant biotechnology are the result of several factors. They include a GM-wary European public, a precautionary political culture when it comes to environmental risks, the weak legitimacy of the European Commission, and decision-making procedures that entail high thresholds of member state consensus to authorise GM products or to change GM product standards.

Breaking the log jam through regulatory change?

One recent development, and another in the proposal stage, provide insight into two attempts to get out of the EU GMO risk regulation log jam. The first represents a modest effort to harmonise EU GM imported animal feed standards with those of exporting countries. The second is a more radical initiative that would give member states greater flexibility to allow, restrict, or ban the cultivation of GM crops in their territory.

GM animal feed imports

Most EU farmers who raise livestock (cattle, hogs, and poultry) rely on imported feed for their animals during the winter months. As their major feed suppliers in North and South America increasingly produce such feed from GM soybeans and corn, EU livestock farmers – particularly in Ireland, the United Kingdom, the Netherlands, Spain, and Portugal – have found it more expensive to source the less available non-GMO feeds. When they turn to GM feed sources, they also face the problem of the EU's zero tolerance policy on imports of GM feeds that have not been approved for sale in the EU.[6] This prohibition against the import of GM products containing even trace amounts of GM strains not yet approved for commercialisation in the EU translates into scarce feed supplies and additional feed costs for European livestock farmers as they wait for the EU's slow regulatory approval process to catch up with that of GM feed exporters. In a joint statement issued in October 2009, organisations representing European grain traders (COCERAL), feed manufacturers (FEFAC), farmers (COPA–COGECA), the oil and protein industry (FEDIOL) and biotechnology companies (EuropaBio) argued that feed costs were three and a half to five billion Euros higher as a result of the EU's zero tolerance policy (AgraFocus August 2009). Those costs were incurred when a large number of shipments of feed containing non-EU authorised GM material were rejected at EU ports in 2009.

In place of a zero tolerance policy for the adventitious presence of unapproved GM varieties in GM feed, the above coalition and DG Agriculture and Rural Policy proposed a low level presence (LLP) policy that allows trace amounts of unapproved genetically modified material in imports of animal feed but not human food. Obtaining agreement on a LLP has been a protracted process. The pan-European environmental lobby and consumer organisations have been implacably opposed to any kind of relaxation in Europe's anti-GM stance, and some member states, including France, opposed any relaxation of the zero tolerance rule without a full risk assessment of the imported product by EFSA. By spring 2011, political leaders in a majority of member states had agreed to relax the zero tolerance policy (EurActiv 2009b and 2009c) in favour of a LLP policy. Their decision, expected to become a regulation by the end of May 2011, allows for import into the EU of animal feed containing up to 0.1 per cent GM material, and which has been approved in a non-EU producing country but not yet in the EU. Prior to importing the feed, EFSA must have conducted a risk assessment to confirm that the unapproved GM material in the feed is not detrimental to health and environment. Moreover, a request to authorise the GM product (animal feed) must have been placed before EFSA for at least three months. Member states do have the right to adopt emergency measures to limit imports of the animal feed if it is shown to constitute a serious risk to human health, animal health, or the environment. The Commission could also suspend all imports in such an event (AgraFocus March 2011a: 22; EurActiv 2011). The LLP policy applies only to an adventitious presence of up to 0.1 per cent of non-authorised GM material in animal feed. The fact that it does not apply to imports destined for human consumption (i.e. food) has elicited criticism from farm groups, the EU oil and protein–meal industry, and GM grain and oilseed exporters who warn that it is impractical and costly to separate global grain supplies into those destined for humans and those for animals (AgraFocus March 2011a: 22).

Member state agreement to allow into the EU animal feed containing GM material that has undergone a risk assessment, but not yet been authorised under EU regulatory procedures, is an important development. A qualified majority of member states approved the LLP policy in contrast to their failure to act coherently on most of the other GM product authorisations. They did so in recognition that the economic interests of European farmers were imperilled by a continuing failure to ignore market developments with respect to GM products outside the EU.

Devolving regulatory authority over GM crop cultivation to member states

A second development, if it comes to fruition, could also potentially break the logjam in the authorisation of GM crops for cultivation. As discussed earlier in this chapter, member states have been highly sensitive to the fact that some of their fellow EU members do not want GM crops to be grown in their country. That sensitivity has left the Commission (and the EU as a whole) in violation of the WTO 2006 panel ruling in *EC-Biotech*, which stated that member state safeguard bans were illegal (without scientific evidence to support them). These national bans are also inconsistent with the principles of the single EU market, which require member states to avoid measures that disrupt the free movement of goods and services within the European Unity. Such a situation, EU Health (SANCO) Commissioner John Dalli has concluded, is not tenable (AgraFocus November 2010: 35).

The 'workable and legitimate exit' that Commissioner Dali is proposing out of this impasse is to give 'wide room for manoeuvre to the member states' in deciding whether to allow the cultivation of GM crops (as quoted in AgraFocus November 2010: 35). A proposal to do so was first floated by the Dutch government in February 2009, endorsed at the June 2009 Environment Council by Austria with the support of twelve other member states, and backed by the Commission in March 2010 (Commission 2010; Council 2009; EurActiv 2010; GMO Compass 2009).

By March 2011, the main elements of a proposal for a regulation that would amend the existing legislation on the deliberate release of GMOs into the environment (Directive 2001/18) were clear. EFSA would continue to be responsible for conducting the health and environmental risk assessment of GMO applications, including setting an appropriate level of protection (ALOP) for a GMO application. In addition, the existing GMO authorisation procedures would also continue; that is, GMOs would be authorised for commercialisation across the EU member states. Thereafter, however, member states would be able to 'opt out' of EU authorisation and restrict or ban a GMO on grounds other than health and/or environmental protection. A working document published by DG SANCO in February 2011 included ethical, moral, and cultural grounds as potential reasons for member states to implement 'proportionate' and 'non-discriminatory' restrictions on GM crops (AgraFocus, March 2011b: 25). In response to member states' request for an indicative list of justifiable grounds to restrict GM crops, the Commission has proposed, beyond moral and ethical reasons, the following: to preserve certain habitats and eco-systems as well as traditional farming methods, and to prevent unintended traces of GMOs getting into other products (AgraFocus March 2011b: 25).

Not surprisingly, the proposal has been controversial, as member states have weighed its implications for the internal market, and its compatibility with both the internal market and WTO rules. Some member states have been strongly opposed, most notably France and Germany, on the grounds the proposal would fragment the internal market. Others question whether it would work in practice and whether it is compatible not only with internal market rules but also with WTO rules. Against these critics, other member states have been more positive. They include Austria, which called in July 2009 for a re-nationalisation of GMO cultivation authorisation to give member states the right to ban GMOs for cultivation on the basis of biodiversity protection considerations and socio-economic criteria. The proposal also enjoys support from the Netherlands, Austria, Hungary, Greece, and Baltic member states (AgraFocus October 2010: 2–3), while the support of other EU states appears to be contingent on assurance that the proposal is consistent with EU Treaty obligations regarding the internal market as well as with WTO law. Interestingly, these member states lend support to the proposal of the Commission for divergent reasons. The Netherlands hopes to be able to develop and experiment with GM crops and thus seeks to circumvent the blockage created by member states such as Austria and Hungary by assuming direct responsibility for its own decisions.

For its part, the European Parliament has been supportive of the proposed regulation, arguing that EU countries should be able to restrict or opt out of cultivating GM crops in their territory on both environmental and socio-economic grounds (European Parliament 2011).

On the issue of the proposal's legality, the legal services of the European Council and the European Commission have diverged. While the Council's legal service have expressed the view that the proposal is contrary to the obligation in Lisbon Article 114 to ensure a more efficient functioning of the single market (AgraFocus December 2010b: 36), the Commission legal services disputes this view. In its judgement, the proposed regulation will make it possible for the internal market to function smoothly, something that is not happening with national safeguard clauses adopted to address concerns unrelated to health and/or environmental protection (AgraFocus December 2010c: 37).

Regarding the compatibility of the proposed regulation with WTO rules, the US Trade Representative has publicly stated his apprehension that the proposal will result in 'Member States all coming up with their own rules' – an outcome that is inconsistent with 'an open transparent process [of GMO authorisation] that conforms with internationally accepted scientific standards' (as quoted in AgraFacts August 2010: 19).

At the time of writing, the fate and eventual substance of the proposed regulation to allow EU countries to restrict or ban GM crops on grounds other than their demonstrated risks to the environment or health is unknown. If it does indeed become law, it seems clear that current bans on GMOs in some countries will continue. Yet it is also possible that in some EU countries – those where public opinion is less hostile to GMOs – farmers would begin to grow GM crops. In general, industry is not happy with this solution because it fears that it will create, to some extent, further legal uncertainty and also divide the farming community in the EU. Since the Commission has no plans to manage competing or diverging authorisation for planting of GM seeds, industry is afraid that the proposed compromise will be of little help in breaking through the deadlock concerning the planting of GM seeds, and instead will continue to sow a hostile and uncertain climate for GM seed importers/manufacturers and farmers (ESA 2010).

Stability in regulatory framework

Finally, the retail market for food products containing GM-derivative ingredients remains unchanged and closed to biotech foods. Although most shoppers in controlled experiments do not actively avoid GM-labelled products, retail stores refuse to carry products with labels indicating the presence of GMO ingredients (European Commission 2008b). There are no GM food products available in European supermarkets. Retail stores are the gatekeepers and many have stated their policies of excluding GM ingredients from their own label products. International food companies apply the same standards to branded items, resulting in a situation where hardly any packaged foods are sold in the European market containing GMO-derived ingredients. According to a 2008 in-depth multi-country survey, only one GM-derived product is sold in the retail market of a dozen member states (Belgium, the Czech Republic, Estonia, France, The Netherlands, Poland, Slovakia, Spain, and the UK); namely, GM-soy oil or lecithin derived from GM-soybean and packaged in cooking oil or margarine (Moses 2008; Moses and Fischer 2008).

Moreover, both the food retailers and some European governments exceed EU requirements in responding to GM-wary consumers. EU legislation does not require meat, dairy, and eggs derived from animals fed with GM animal feed to be labelled. However, both France and Germany have allowed GMO-free labelling for these products. French rules provide for a 'GMO-free' label if these products come from animals who have consumed feed with a GM content less than 0.1 per cent rather than the EU threshold of 0.9 per cent. German 'GM-free' logo can be applied to foods derived from animals where no GM plants were used in the feed.

Thus, the retail market for foods containing GM ingredients will continue to be closed. Moreover, in contrast to the turmoil surrounding the import of GM feed and planting of GM seeds, there are few demands or pressures to revisit the labelling rules imposed by Regulation (EC) 1830/2003 (concerning the traceability and labelling of GMOs). Food processors have made the necessary adjustments and have been able to supply the European food retail markets with food products free of GM ingredients. Thus far, there are no external pressures or internal demands that would result in new political negotiations and possibly new EU rules.

Conclusion

In this chapter, we have discussed the EU risk regulation regulatory framework, providing an explanation for why it has taken the form it has and also why regulatory procedures to authorise GMOs have been bedevilled by controversy and largely ineffective. This explanation has highlighted the impact on EU GMO policy of, first, a negative European public opinion when it comes to GM foods; and second, an institutional decision-making framework that requires a high threshold of agreement among political actors who diverge significantly in their views of the desirability of GM products and how best to regulate them. While it is internal factors in the EU that overwhelmingly explain regulatory policy developments, the latter have also been affected by factors external to the EU, including WTO rules (Skogstad 2011) and the widespread cultivation of GM crops across the globe, including in countries on which the EU is reliant for imported feed.

We have also discussed recent initiatives in EU GMO regulatory developments that we believe represent efforts to break the logjam in GMO authorisations in the EU. The proposal to allow member states to opt out of EU level GMO authorisations and restrict GM products in their countries is clearly significant. Whether it will result in a more GMO-friendly Europe is another matter. Polling data show that Europeans' aversion to GM foods is not rooted in a lack of familiarity with GM foods or in scientific ignorance; being more familiar with GM foods does not lead to more positive perceptions of it, nor are those with a scientific background or science degree more likely to support GM foods (European Commission. Directorate-General for Research 2010: 100). As long as these public views prevail, the European food retail market will continue to be resistant to food derivatives based on genetically engineered products.

In retrospect, it would appear that agricultural biotechnology should have remained under the control of national authorities and domestic institutions because the EU member states fundamentally disagree about the utility of biotechnology, about frameworks to establish risk assessment, and about the future of genetic engineering. At first, EU officials figured that stringent authorisation rules and elaborate decision-making procedures would reassure national officials that the EU had taken all precautionary and deliberate measures to evaluate potential risk factors. However, the complex decision-making structure itself has not silenced critics of genetic engineering; instead, it has become a major source of friction, both within the EU and between the EU and the US and other farming regions in the Americas. Realising that the decision making procedures do not work because they assume a level of consensus and agreement that is absent, the EU is now willing to contemplate the re-nationalisation of the cultivation of GM seeds in order to solve the transatlantic conflict as well as competing visions in the EU.

Whether re-nationalisation of GM planting decisions will be a viable solution is to be seen. Delegating new authority to national governments so that national politicians can make their own decisions with respect to the planting of GM seeds will inevitably result in divergent rules and standards. In turn, the presence of different national rules challenges the integrity of the single market and will not necessarily lift the current blockage on GM cultivation. Member states can still invoke the safeguard clause, and ban the planting of food/feed in their fields. Since seeds may travel to adjacent areas in other member states or regions, governments in member states with safeguard clauses would rule the presence of traces of the GMO as illegal. For farmers and seed manufacturers, this possibility causes great uncertainty and may discourage them from planting GM seeds in the first place. Since most GM seeds are sold by American companies, it is expected that the standoff between the EU and the US will continue for the foreseeable future and remain a source of friction and conflict.

Notes

1 The pieces of legislation are Directive 2001/18, the Deliberate Release Act; Regulation 1829/2003, the Food and Feed Regulation; Regulation 1830/2003 on the labelling and traceability of GM products; and Regulation 178/2002 which is not specific to GM products but which lays out the role of the European Food Safety Authority.
2 The criticism of inadequate scientific risk assessments prior to approving GMOs related to the fact that the European Commission had discretion on whether to consult scientific advisors.
3 Austria has had a ban on GM crops since 1999 while the other national bans are more recent and usually specific to a single product. Greece's ban on MON810 came in April 2005, Poland's in May 2006, Hungary's in September 2006, France's in February 2008, Romania's in March 2008, and Germany's in April 2009. In March 2011, Bulgaria announced that it also would ban MON810. Italy invoked a general ban on cultivation of all GM crops in March 2006 until it implemented legislation to allow the coexistence of conventional and GM crops. Hungary and Luxembourg also banned the GM potato following its approval in 2010. This information was accessed from GMO-Free Europe 2010 on 23 April 2011 at http://www.gmo-free-regions.org/gmo-free-regions/bans.html.
4 Spain is by far the largest cultivator of Bt maize in the EU.
5 See note 3.
6 The European Commission's Directorate for Agriculture and Rural Affairs forewarned in a 2008 report of the problem of imported feed costs of non-GM feed, as well as that of traces of unapproved GM varieties in feed shipments from the US.

Bibliography

AgraFacts (August 2010) 'USTR Ron Kirk Voices Concern About EU Proposals on GM Cultivation Rules': 19. Brussels.

AgraFocus (June 2006) 'EFSA promises greater transparency on GMOs': 36. Brussels: AgraEurope.

AgraFocus (November 2007) 'Dimas Stance on GMO Approvals Leaves Commission Split': 35. Brussels: AgraEurope.

AgraFocus (September 2008) 'GMO *Ad Hoc* Group to Tackle 5 Broad Themes': 24. Brussels: AgraEurope.

AgraFocus (August 2009) 'Loss of US soya suppliers could cost EU industry €3.5–5 billion': 29–30. Brussels: AgraEurope.

AgraFocus (August 2010) 'USTR Ron Kirk Voices Concern About EU Proposals on GM Cultivation Rules': 19. Brussels: AgraEurope.

AgraFocus (October 2010) 'GMO Discussion: Member States divided on GMO proposals; Germany leads opposition': 2–3. Brussels: AgraEurope.

AgraFocus (November 2010) 'ENVI Ministers Challenge Commission GMO Proposals': 35. Brussels: AgraEurope.

AgraFocus (December 2010a) 'EFSA provides guidelines on ERA': 39. Brussels: AgraEurope.

AgraFocus (December 2010b) 'Commission GM Plans Legally Flawed, Says Council': 36. Brussels: AgraEurope.

AgraFocus (December 2010c) 'Commission Legal Team Refutes Doubt Over Legal Basis of GMO Package': 37. Brussels: AgraEurope.

AgraFocus (March 2011a) 'Green Light to Commission Measures to Address GMO "Zero Tolerance"': 22. Brussels: AgraEurope.

AgraFocus (March 2011b) 'GMO working party demands legal clarity on opt-outs': 25. Brussels: AgraEurope.

Ansell, Chris and David Vogel (2006) 'The contested governance of European food safety regulation', in Chris Ansell and David Vogel (eds) *What's the Beef?: The Contested Governance of European Food Safety*. Cambridge, MA: MIT Press.

Bauer, M. W. and G. Gaskell (2002) *Biotechnology: the making of a global controversy*. New York: Cambridge University Press.

Bernauer, Thomas (2003) *Genes, Trade, and Regulation: The Seeds of Conflict in Food Biotechnology*. Princeton: Princeton University Press.

Bernauer, Thomas and P. Aerni (2007) 'Competition for Public Trust: Causes and Consequences of Extending the Transatlantic Biotech Conflict to Developing Countries', in Robert Falkner, (ed.) *The International Politics of Genetically Modified Food: Diplomacy, Trade and Law*. New York: Palgrave Macmillan.

Bernauer, Thomas and Ladina Caduff (2004) 'In Whose Interest? Pressure Group Politics, Economic Competition and Environmental Regulation', *Journal of Public Policy*, 24: 99–126.

Bonneuil, Christophe, Pierre-Benoit Joly and Claire Marris (2008) 'Disentrenching experiment: The construction of GM-Crop field trials as a social problem', *Science Technology and Human Values*, 33: 201–29.

Borrás, Susana (2006) 'Legitimate governance of risk at the EU level? The case of genetically modified organisms', *Technological Forecasting & Social Change*, 73: 61–75.

Borrás, Susana (2007) 'Governance Networks in the EU: The Case of GMO Policy', in M. Marcussen, and J. Torfing (eds) *Democratic Network Governance in Europe*. Houndmills: Palgrave Macmillan.

Buonanno, Laurie (2006) 'The creation of the European food safety authority', in Chris Ansell and David Vogel (eds) *What's the Beef?: The Contested Governance of European Food Safety*. Cambridge, MA: MIT Press.

Commission of the European Union (2010) Proposal for a REGULATION OF THE EUROPEAN PARLIAMENT AND OF THE COUNCIL amending Directive 2001/18/EC as regards the possibility for the Member States to restrict or prohibit the cultivation of GMOs in their territory (COM(2010) 375 final).

Council of the European Union (2009) Council Decision concerning the provisional prohibition of the use and sale in Austria of genetically modified maize (6330/09). Brussels 11 February 2009. Available at http://register.consilium.europa.eu/pdf/en/09/st06/st06330.en09.pdf.

EFSA (2008) 'Request from the European Commission to Review Scientific Studies Related to the Impact on the Environment of the Cultivation of Maize Bt11 and 1507: Scientific Opinion of the Panel on Genetically Modified Organisms', *EFSA Journal* 851: 1–27.

ESA (2008) *European Commission's orientation debate on GMOs confirms existing rules and procedures – but fails to show leadership on the future policy*. Brussels: ESA_08.0308. (7 May 2008).

ESA (2010) *European Commission's proposals for a new policy approach to the authorisation of GMOs for EU cultivation and coexistence of GM and non-GM crop production*. Brussels: ESA_10.0546.2 (23 June 2010).

EurActiv (2009a) 'Ministers back right to refuse GM crop cultivation', (March 3, 2009) http://www.euractiv.com/en/sustainability/ministers-back-right-refuse-gm-crop-cultivation/article-179882.

EurActiv (2009b) 'EU farm chief pushes for biotech feed rules', 15 September 2009. http://www.euractiv.com/en/cap/eu-farm-chief-pushes-biotech-feed-rules/article-185448.

EurActiv (2009c) 'Crisis looming as EU blocks GM-soy imports', 23 October 2009. Document4http://www.euractiv.com/en/cap/crisis-looming-eu-blocks-gm-soy-imports/article-186681.

EurActiv (2010) 'EU to overhaul GM crop approval system'. Available at http://www.euractiv.com/cap/eu-overhaul-gm-crop-approval-system-news-494896.

EurActiv (2011) 'EU experts approve trace GM in feed imports: Official', 23 February 2011 http://www.euractiv.com/en/cap/eu-experts-approve-trace-gm-feed-imports-official-news-502442.

EuropaBio (2007) *Commissioner Dimas' Rejection of Two GM Maize Products Would Undermine the EU's Safety Assessment Process*. Brussels: EuropaBio (27 November 2007).

European Commission (17 July 1999) 'Council decision of 28 June 1999 laying down the procedures for the exercise of implementing powers conferred on the Commission', (1999/468/EC) *Official Journal* L 184: 23.

European Commission (2001) 'Directive 2001/18/EEC of the European Parliament and of the Council of Ministers of 12 March 2001 on the deliberate release into the environment of genetically modified organisms and repealing Council Directive 90/220/EEC', *Official Journal of the European Communities*. L106/1-39,17/04/2001. Available at: «http://eur-lex.europa.eu/pri/en/oj/dat/2001/l_106/l_10620010417en00010038.pdf».

European Commission (2003) 'Regulation (EC) No. 1830/2003 of the European Parliament and of the Council of 22 September 2003 concerning the traceability and labelling of genetically modified organisms and traceability of food and feed products produced from genetically modified organisms and amending Directive 2001/18/EC(16)',*Official Journal of the European Union*. L268/24-28, 18.10.2003. Available at: «http://eur- lex. europa.eu/pri/en/oj/dat/2003/l_268/l_26820031018en00240028.pdf».

European Commission (2008a) 'European Commission Discusses How Best to Move Forward on the Authorisation of Pending GMO Applications', No. 44/08 (Washington, DC: Delegation of the European Commission to the USA) (7 May 2008).

European Commission (2008b) *Do European Consumers Buy GM Foods?* London: King's College London ('Consumerchoice') http://www.kcl.ac.uk/consumerchoice.

European Commission Directorate-General for Research (2010) *Europeans and biotechnology in 2010; winds of change?* Luxembourg. Available at http://ec.europa. eu/public_opinion/archives/ebs/ebs_341_winds_en.pdf.

European Parliament (2011) 'EU countries should be able to ban GMOs on environmental grounds-MEPs', downloaded from the European Parliament website on 23 April 2011: http://www.europarl.europa.eu.

European Union, (08/05/1990) 'Council Directive 90/220/EEC of 23 April 1990 on the deliberate release into the environment of genetically modified organisms', *Official Journal* L 117: 0015–27.

Everson, Michelle and Ellen Vos (eds) (2009) *Uncertain Risks Regulated: Facing the Unknown in National, EU and International Law*. New York: Routledge.

Falkner, Robert (2007) *The International Politics of Genetically Modified Food: Diplomacy, Trade and Law*. New York: Palgrave Macmillan.

Ferretti, Maria Paola and Pavone, Vincenzo (2009) 'What do civil society organizations expect from participation in science? Lessons from Germany and Spain on the issue of GMOs', *Science and Public Policy*, 36: 287–99.

Friends of the Earth Europe (2004) *Throwing Caution to the Wind: A Review of the European Food Safety Authority and its work on Genetically Modified Foods and Crops*. Available at: http://www.foeeurope.org/GMOs/publications/EFSAreport.pdf.

Gaskell, George, Paul Thompson and Nick Allum (2001) 'Worlds Apart? Public opinion in Europe and the USA', in George Gaskell and Martin W. Bauer, (eds) *Biotechnology 1996–2000*. London: Science Museum (351–75).

GMO Compass (2009) 'EU countries should be able to ban genetically modified plants'. Available at http://www.gmo-compass.org/eng/news/453.eu_countries_should_able_ban_genetically_modified_plants.html.

GMO Compass (2010) 'EU Commission: Import approval for six GM maize lines', (28 July 2010) Document4http://www.gmo-compass.org/eng/news/526.docu.html.

Gottweis, Herbert (1998) *Governing Molecules: The Discursive Politics of Genetic Engineering in Europe and the United States*. Cambridge, Mass.: MIT Press.

Heller, Chaia (2002) 'From Scientific Risk To Paysan Savoir-Faire: Peasant Expertise in the French and Global Debate over GM Crops', *Science as Culture*, 11: 5–37.

Hu, W., M. M. Veeman and W. L. Adamowicz (2005) 'Labelling Genetically Modified Food: Heterogeneous Consumer Preferences and the Value of Information,' *Canadian Journal of Agricultural Economics/Revue canadienne d'agroeconomie*, 53: 83–102.

ISAAA (International Service for the Acquisition of Agri-biotech Applications) (2011) *Global Map of Biotech Crop Countries and Mega-Countries in 2010*. Available at: www.isaaa.org.

Isaac, G. (2002) *Agricultural Biotechnology and Transatlantic Trade*. Oxford: CABI Publishing.

Jasanoff, S. (2005) *Designs on Nature: Science and Democracy in Europe and the United States*. Princeton, NJ: Princeton University Press.

Kurzer, Paulette (2011) 'Beyond the welfare state: Consumer Protection and Risk Perceptions in the European Union and Austria', in Robert Cox and G. B Cohen (eds) *Social Policy in the New Europe: The Experience of the Smaller EU Members*. New York: Berghahn Books.

Kurzer, Paulette and Alice Cooper (2007) 'What's for Dinner? European Farming and Food Traditions Confront American Technology', *Comparative Political Studies*, 40: 1035–58.

Levidow, Les (2006) 'EU Agbiotech Regulation', *Soziale Technik*, 3: 10–2.

Levidow, Les (2007) 'The Transatlantic Agbiotech Conflict: A Policy Problem and Opportunity for EU Regulatory Policies', in Robert Falkner (ed.) *The International Politics of Genetically Modified Food: Diplomacy, Trade and Law*. New York: Palgrave Macmillan.

Meins, Erika (2003) *Politics and Public Outrage: Explaining Transatlantic and Intra-European Diversity of Regulations on Food Irradiation and Genetically Modified Food*. London: LIT Verlag.

Moses, Vivian (2008) 'Discussion and Conclusion', *Do European Consumers Buy GM Foods?* London: King's College http://www.kcl.ac.uk/consumerchoice.

Moses, Vivian and Siglinde Fischer (2008) 'Introduction', *Do European Consumers Buy GM Foods?* London: King's College http://www.kcl.ac.uk/consumerchoice.

Noussair, Charles, Stephane Robin and Bernard Ruffieux (2002) 'Do consumers not care about biotech foods or do they just not read the labels?', *Economics Letters* 75: 47–53.

Pollack, M. A. and G. Shaffer (2009) *When Cooperation Fails: The International Law and Politics of Genetically Modified Foods*. New York: Oxford University Press.

Ramjoué, Celina (2007) 'The transatlantic rift in genetically modified food policy', *Journal of Agricultural and Environmental Ethics*, 20: 419–36.

Seifert, Franz (2010) 'Back to Politics at Last Orthodox Inertia in the Transatlantic Conflict over Agro-Biotechnology', *Science, Technology & Innovation Studies*, 6: 101–26.

Skogstad, Grace (2003) 'Legitimacy and/or Policy Effectiveness? GMO Regulation in the European Union', *Journal of European Public Policy*, 10(3): 321–38.

Skogstad, Grace (2011.) 'Contested Accountability Claims and GMO Regulation in the European Union', *Journal of Common Market Studies*, 49(4): 1–21.

Tiberghien, Yves (2009) 'Competitive governance and the Quest for Legitimacy in the EU: the Battle over the Regulation of GMOs since the mid-1990s', *Journal of European Integration*, 31(3): 389–408.

Toke, D. (2004) *The Politics of GM Food: A Comparative Study of the UK, USA, and EU*. London: Routledge.

Vogel, David (1995) *Trading Up: Consumer and Environmental Regulation in a Global Economy*. Cambridge: MA: Harvard University Press.

Young, Alasdair R. (2009) 'Confounding Conventional Wisdom: Political not Principled Differences in the Transatlantic Regulatory Relationship', *British Journal of Politics & International Relations*, 11: 666–89.

Young, Alasdair R. (2010) 'Of executive preferences and societal constraints: The domestic politics of the transatlantic GMO dispute', *Review of International Political Economy*, 18: 1–24.

Winickoff, David, Sheila Jasanoff, Lawrence Busch and Robin Grove-White (2005) 'Adjudicating the GM food wars: science, risk, and democracy in World Trade Law', *Yale Journal of International Law*, 30: 82–121.

Part 4

Lessons from other high technology sectors

9 Regulatingn anotechnology in China

Governance, risk management, and regulatory effectiveness[1]

Darryl S. L. Jarvis and Noah Richmond

Introduction

Our ability to arrange atoms, observes K. Eric Drexler, lies at the foundation of technology (Drexler 2006: 55). It is merely variation in the arrangement of atoms that differentiates sand from computer chips, cancer from healthy tissue, and gold from bauxite. Now imagine a series of technologies that change the molecular structure of biological entities, proteins, DNA, and the building blocks that generate and control biological outcomes. Or, imagine a series of technologies that are capable of engineering molecular and atomic variation in the composition of compounds to produce new materials with new properties and characteristics. Imagine further, as Drexler does, that 'DNA engineering builds precise, million-atom frameworks', such that 'engineered proteins can bind to precise locations on these frameworks', and 'proteins can bind other components' that are electrically or chemically active so that these proteins and the biological structures on which they are attached can 'serve as construction machinery' (Drexler 2006:12). If you can imagine all this, you can imagine *nanotechnology*: a diverse collection of academic specialisms centered around engineering and manipulating molecular and atomic structures and, in the process, creating biological and non-biological nanomaterials whose characteristics can be 'made to order.' Penultimately, imagine self-replicating nanobots able to organize the atomic outcomes we desire, carrying out molecular level construction and repair work.

Nanotechnology is the science of manipulating matter at the atomic and molecular scale. It deals with structures sized from 1–100 nanometers in dimension – one nanometer being equal to one billionth of a meter (Renn and Roco 2006: 153; see also Lindquist, Mosher-Howe and Lui 2010). This, in a sense, is God in a tea cup; it holds vast prospects for technological innovations in areas such as: electronics, through the development of nanocircuitry; molecular level semiconductors; nanotubes; new materials development in ceramics, polymers, glass ceramics and composites; and in medicine, through the development of nano-electronic biosensors, nano-scale drug particles, and delivery systems to improve the accuracy and efficiency of drug toxicity to harmful tissue and disease – among many others.

Far from science fiction, nanotechnology and the development of nanomaterials is already well advanced. Nanomaterials are currently present in over 1,200 commonly consumed products ranging from cosmetics, clothing, personal care/ hygiene items, sporting goods, sunscreen, and household filtration systems and construction materials. If you consumed a McDonald's hamburger recently, the paper ring that held the burger together was glued with a nano-based resin; if you had a wound dressed with an "Acticoat" dressing or applied "Acnel" lotion to dry skin, then each of these products had nanomaterials incorporated into their production (Project on Emerging Nanotechnologies 2010). The US National Science Foundation estimates that US$70 billion worth of nano-containing items are sold in the United States each year, while the global market for manufactured goods containing nanotechnology is estimated to reach US$2.6 trillion by 2014 (AOL 2010; see also Lux Research). Moreover, the rate of development and incorporation of nanotechnologies into all facets of consumer, industrial, and medical applications is anticipated to double every two years.

Science and nanotechnology in China[2]

While the United States has led global investment into research and development in nanotechnology, China is fast emerging as a global player. By 2005, China ranked only second to the US in nanotechnology investment, ranked second in terms of the number of nano-related peer reviewed research publications (producing fifteen percent of all global nano-related research papers), and had emerged as the global leader in carbon nanotube technology and manufacture. China is also a leader in the manufacture of nano coatings, anti-corrosive nano paints (used in ship construction and oil tanks), odor eating nano coatings and plastics for refrigerators, nano filters for air conditioners, a series of nano materials used in optics (to filter glare), and in the production of nano textiles and clothing (to enhance antimicrobial properties) (Shapira and Wang 2009: 461; see also Liu and Zhang 2007).[3]

China's push to become a global leader in nanotechnology reflects a national strategy aimed at leapfrogging the developmental cycle. While development of the export sector has facilitated rapid economic growth, primarily through specialization in low to medium value-adding manufacturing, sustained growth will be contingent on moving up the value chain. Leading Chinese policymakers, economic planners and influential economists all recognize the need to address China's dependence on export-led growth. As the Vice President of the China National Academy of Nanotechnology and Engineering (CNANE) notes, 'China must break away from the mode of technology dependence and transform into independent technology innovation … It is very clear that [in China] the leading power is in the tight grasp of foreign enterprises' (as quoted in Appelbaum and Parker 2008: 319).

China's science and technology policy is thus informed by a singular rationale: economic growth situated in the context of developing indigenous scientific and technological capacity to reduce reliance on technology transfer, export-led growth, and low end manufacturing. Importantly, China sees its science and technology policy as a central pillar in its efforts to become a global leader in innovation; that is, a net exporter of ideas, innovative technologies, and commercial applications.

Developing nanoscience and nanotechnology in China: policy and regulatory frameworks, 1990–2020

Chinese domestic spending on science and technology (S&T) research and development has exploded since 1996, growing ten-fold from RMB 40 billion to RMB 461 billion in 2008. As a proportion of gross domestic product (GDP), China's investment into S&T research grew from 1.5 percent in 1996 to 2 percent in 2010, and, under the Medium and Long Term Development Plan (MLP) running from 2006–20, it will grow to 2.5 percent of GDP. Yet despite such significant increases in the level of national investment into S&T, China still trails countries such as South Korea and Japan whose technology spending as a share of GDP stands at 3.5 percent and 3.4 percent, respectively.

China's science and technology programs are situated around a central policy architecture announced by Deng Xiaoping in 1986, the National High Technology Research and Development Program, known as the 863 Program. The 863 Program aims at 'promoting the development of key novel materials and advanced manufacturing technologies for raising industry competitiveness' including nanomaterials. The 863 Program is implemented through successive Five-Year Plans and is the key government program behind the national research and development (R&D) capacity in support of domestic innovation. Indeed, from 1990–2002 the 863 Program funded over 1,000 nanotech projects with a total investment of US$ 27 million. Similar to the MLP, the Chinese government has recognized the importance of linking S&T to economic growth through targeted research projects. To help achieve this, the 863 Program is managed by an expert responsibility system, with field/sector specific expert committees and panels consisting of top scientists who supervise, advise, and assess projects.

The first project adhering to the 863 Program goals was the Climbing Project on Nanomaterial Science instigated from 1990–9 and overseen by the State Science and Technology Commission (SSTC), the predecessor to the current Ministry of Science and Technology (MOST). Given the Program's success, the government subsequently renewed its commitment to funding basic research on nanomaterials and nanostructures (i.e. carbon nanotubes) with the initiation of China's National Basic Research Program (973 Program) in 1997. This complements the 863 Program and is an evolving research agenda for nanotechnology research. Since 2006, ten nanotechnology research projects have received a combined US$ 30 Million (US$ three million each) under the Program.

The 973 Program also supports fundamental research that enhances domestic capacity and knowledge on nanotechnologies while complementing the 863 Program in applied research. Two other notable projects under the 973 Program concern the standardization of procedures and assessment/test protocols, which form the basic framework for the regulation of nanomaterials. The first, the standardization for the key measurement techniques in nanotechnology is led by Professor Jiang Chao at the National Center for Nanoscience and Technology (NCNST), and the second, the Controlled Synthesis of Nanometer-sized Reference Materials for Metrology and Measurement: Scaling-up and Standardization of Nanofabrication Methods, is led by Professor Wu Xiaochun, also at the NCNST. Each is a standards-based series of protocols designed for benchmarking nanoscience research and findings.

In addition to nanotechnology research funding, the tenth Five-Year Plan (2001–5) also addressed priorities for the commercialization and development of nanotechnology. The government disaggregated nanotechnology development between short (development of nanomaterials), medium (development of bio-nanotechnology and nano medical technology), and long term projects (development of nano electronics and nano chips). The Five-Year Plan prioritized bridging the gap between nanotechnology research and market demand to form a complete national innovation system. The eleventh Five-Year Plan (2007–12) in turn places emphasis on innovative technologies, including the development of new materials for information, biological, and aerospace industries, and commercializing the technology for 90-nanometer and smaller integrated circuits.

As part of China's longer term S&T policy objectives, the MLP 2006–20 is a follow-up to the Five-Year Plan, and designed to provide China with the necessary technical capacity for sustained technology innovation that contributes to national economic development and China's ambitions to become a global leader in S&T research. Indeed, the MLP 2006–20 plan calls on S&T to address current development bottlenecks such as environmental degradation and energy efficiency and thus avert negative externalities associated with accelerated growth. Importantly, under the MLP, nanotechnology development is given priority status and is identified as one of science's 'megaprojects.' It calls for R&D to be undertaken on nanomaterials and devices, design and manufacturing technology, nano-scale complementary metal-oxide semiconductor devices, nano drug carriers, energy conversion and environmental purification materials, and information storage material. Between 2006 and 2008, the MLP funded twenty-nine nanotechnology projects in twenty-two universities and research institutes across the country, totaling US$ 38.2 million.

To help oversee the various nano projects, the National Steering Committee for Nanoscience and Nanotechnology (NSCNN) was established in 2000 to coordinate and streamline all national research activities. The NSCNN is directed by Dr Chunli Bai and it consists of MOST, the Chinese Academy of Sciences (CAS), National Natural Science Foundation (NSFC), the National Development and Reform Commission (NDRC), the Ministry of Education (MOE) and the Chinese Academy of Engineering (CAE). The NSCNN membership excludes other regulatory bodies such as health, environment, and worker safety ministries. The NSCNN is chaired by the Minister of MOST and includes twenty-one scientists from universities and research institutions and seven officials from government agencies (see Figure 9.1). The preliminary results of our interviews

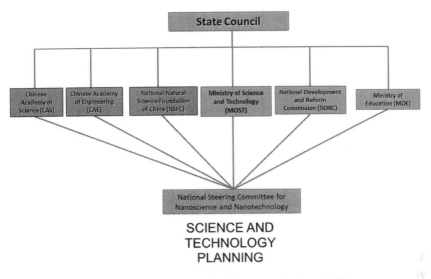

SCIENCE AND
TECHNOLOGY
PLANNING

Figure 9.1 Policy and key regulatory agencies for nanotechnology in China
Source: Interviews 2010.

suggest that the approval of research grants by the NSCNN depends primarily on demonstrating the commercial utility of projects, and that NSCNN is focused predominantly on commercialization objectives.

Managing nano-based risks in China: regulatory responses

As the complexity of nano-based research (pure and applied) has increased in line with increased government funding, and as the number of industrial applications for nanomaterials has grown, China has moved to identify measurement, handling, exposure, toxicity, and safety standards. Nanotechnology standards are reviewed by the National Nanotechnology Standardization Technical Committee (NSTC), the Technical Committee 279, a nanomaterial-specific sub-committee under the Standardization Administration of China (SAC). It is housed within the NCNST. The SAC/TC279 serves as the coordinating body for the purposes of drafting essential nanotechnology standards including terminology, methodology, and safety in the fields of nano-scale measurements, materials, and nano-scale biomedicine.

The NSTC-TC also develops test protocols and technical standards used by manufacturing firms. The Committee oversees applied research for industry and metrology, and laboratory measurement instruments in particular. SAC/TC279 is also constructing a database for nanomaterial toxicology studies to assist in the establishment of safety standards for nanomaterial production, packaging, and transportation. The NSTC-TC has five core research working groups:

1   micro-fabrication,
2   nano-metrology,
3   health, safety, and the environment,
4   nano-indentation testing and,
5   scanning probing microscopy (see Figure 9.2).

Standards are usually published, administered, and enforced by TC279's parent agency, the General Administration of Quality Supervision, Inspection and Quarantine (AQSIQ). Technical standards are distinguished between GB (mandatory), GB/T (voluntary), and GB/Z (technical guide). Of the published nanotechnology standards, seventeen are voluntary. However, during interviews it was evident that the distinction between voluntary and mandatory standards is unclear: complying with voluntary standards often ensures market access for nano products and thus voluntary standards tend to be treated as mandatory by commercial operators. Under this regime, there are multiple standards relating to various nano-related issues. Ten standards, for example, are concerned with new testing methodologies and protocols; five apply to product specifications; and two standardize older terminology. There is only one standard provided by the industry, which is GB/T 22925–2009. It was established by the China Association of Textile Industry (CATI) that provides specifications for nanotechnology-treated clothes (for a complete list of standards see Appendix 9.1 at the end of this chapter).

The regulatory regime for the management of nanotechnology chemicals (still being implemented at the time of writing) will likely manage risks comparable to those identified under the EU's REACH regulation. Under REACH (the

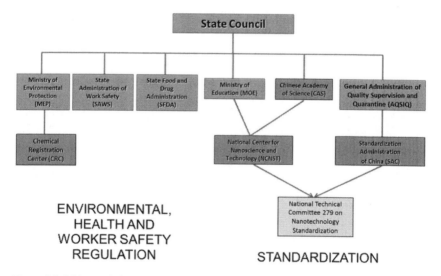

Figure 9.2 Risk regulation of nanoscience and nanotechnology in China
Source: Interviews, 2010.

Registration, Evaluation, Authorization and Restriction of Chemical substances), all chemicals imported or produced in the EU are subject to the same regulatory oversight, requiring companies to submit chemical data to the European Chemical Agency (ECA). Part of the required data includes safe handling guidelines to manage the risks of chemical substances and provision of data on the environmental, health, and safety properties of chemicals.

Presently, chemical regulation in China follows the EU approach in that companies must apply to a central oversight authority, the Chemical Registration Centre of the Ministry of Environmental Protection (CRC–MEP), for a registration certificate prior to manufacture/import of new chemicals. In this application process, companies must supply toxicological and other data to the CRC for all new chemical substances not listed on the Inventory of Existing Chemical Substances Manufactured or Imported in China (IECSC). Registration requirements and enforcement guidelines are set out in the Provisions on the Environmental Administration of New Chemical Substances made effective in September 2003. Penalties for producing chemicals without obtaining registration include paying a fine and being prohibited from registering new substances in China for three years. Falsification of documents and chemical material data during random inspections is also punishable by the same penalties.

In the area of health and safety, China has a complex regime to manage and oversee the use and manufacture of chemical substances. This also applies to nanotechnology with multiple ministries including the Ministries of Health, Communication, Public Security, and multiple administrative organs including the State Food and Drug Administration (SFDA), State Administration of Standardization (SAS), State Administration of Work Safety (SAWS), and AQSIQ – all having some control over environment, health, and safety (EHS) aspects of chemicals. This implies potential coordination difficulties and calls for inter–ministerial/inter–agency coordination to improve the management of potential regulatory gaps. It also makes it harder to establish a clear, predictable, and simplified regulatory environment for companies to operate in.

In 2009, the government revised the chemical substance rules in order to incorporate risk assessment, risk management, and data submission requirements similar to REACH. These requirements include measures to mitigate hazards and exposure in labeling, packaging, transporting, and disposal of new chemicals. The 2009 amendments also require eco-toxicological test data to be drawn from testing organizations at Chinese laboratories. Some concerns have been raised, namely that the technical infrastructure and testing labs might not be of sufficient quality/ capacity to generate timely and OECD-acceptable chemical test data. Yet the 2009 amendments demonstrate the MEP's commitment to adopt the precautionary principle for managing risks. They may also be applicable to nanotechnology since notification will be required for chemical substances manufactured and/or imported under one kilotonne per year, although with the caveat of reduced data submission requirements.

Day-to-day enforcement of chemical production and food safety regulations fall directly onto local governments, though national agencies reserve enforcement power. In all cases, local governments are charged with supervisory and administrative responsibility. This may be problematic when local governments inconsistently implement /interpret national laws (see Figure 9.2).

Nano safety regulation in China: EHS risk

In 2001, CAS researchers recognized the environmental and toxicological effects of manufactured nanomaterials by signaling their concern of the EHS risks of nanomaterials. Because of their findings, the Laboratory for Bio-Environmental Effects of Nanomaterials and Nano-safety (LBENN) was established in 2003. It is currently based at the Institute of High Energy Physics, the CAS, and the NCNST. MOST provides research funding while CAS provides support for capital costs and equipment for EHS research. LBENN employs researchers in the fields of chemistry, biology, nanosciences, and toxicology. In the area of nanosciences, the objectives of LBENN are to first establish a methodology for the detection of nano particles *in vivo* and determine their biological, environmental, and toxicological effects, and then establish protocols for the safe use and handling of nanomaterials and nanotubes for medical and consumer use. These studies are funded by the 973 Program under a five-year nano safety project, namely, the Health and Safety Impacts of Nanotechnology, Exploring Solutions. The studies on EHS risk of nanomaterials also support another objective, which is the development and drafting of regulatory frameworks for research and industrial activities in nanotechnology.

Presently, there are more than thirty research organizations in China that have initiated research activities studying the toxicological and environmental effects of nanomaterials and nanoparticles, and developing techniques for recovering nanoparticles from manufacturing processes (interview 2010). For example, the CAS, Beijing University, and the Chinese Academy of Medical Sciences are conducting toxicology studies on nanomaterials intended for medicines. In our interviews, we were told that a generational divide exists among researchers; younger scientists discount EHS risks and are more motivated towards commercial goals, whereas older researchers take a more precautionary stance. One senior interviewee suggested that safety is not usually an important factor in nanomaterial research, except when it involves medical applications.

In 2006, the SFDA revised its medical device regulations to require medical devices made with nanometer biological materials and medical instruments made with nanometer metal silver material to be classified as Class III medical devices. This means that devices are subject to stricter control and safety oversight than other medical devices, including more detailed pre-market approval procedures. This was the first published case of nano products that were subject to special regulatory requirements.

Given that research funding for nanotoxicology studies is distributed by MOST, and research conclusions are communicated back to the organization, it is not clear how MOST handles its competing objectives of promoting technology development and taking EHS risk research seriously. Driven by a strong need to commercialize research and contribute to economic growth, MOST, and researchers funded by MOST, have strong incentives to underestimate the risks of new technologies.

Regulatory effectiveness and governance gaps

While the full impact of China's push for leadership in nanotechnology and nano-materials is yet to be realized, policy planning, regulation, and management reveal much about the state of the sector, Chinese public attitudes towards nanotechnology, and, in turn, how the discourse and management of possible nano risks are framed and approached by public agencies and regulators. While nanotechnology holds enormous potential for commercial gain, cutting-edge technological innovation, and the development of an innovative knowledge economy, the risks associated with nanotechnologies and nanomaterials on human health and the environment remain largely unknown. Recent laboratory experiments on carbon nanotubes suggest that they could be as dangerous as asbestos fibers (Scientific American 2008; see also Falkner 2008; Scheufele *et al*, 2007). More importantly, nano toxicity is thought to display an inverse relationship to particulate size: the smaller the particulate matter, the more toxic such particulates tend to be (Woodrow Wilson International Center and the Pew Charitable Trusts 2010). However, the precise dimensions of these risks, especially with longer term exposure, or exposure through nanomaterials engineered in chemical composites and utilized in industrial and chemical applications, have yet to be determined. For this reason, nanotech-specific safety regulations, toxicity, and exposure levels have not been formalized, nor a commonly accepted international safety regulatory framework established (Breggin *et al*. 2009).

Science has historically approached new technologies by invoking the 'precautionary principle.' Broadly stated, the precautionary principle assumes that if a technology or policy has a suspected risk of harm (to individuals, the public, or the environment), absent scientific consensus about the extent and magnitude of these risks, the burden of proof that the technology or policy is not harmful, falls to its proponents. In China and the case of nanotechnology, however, the extent to which the precautionary principle guides the adoption of nanotechnologies and the use of nanomaterials appears problematic. Several interrelated factors contribute to this.

First, the discourse framing China's pursuit of nanotechnology is tied intimately to a national political agenda. As one of China's four science-based 'megaprojects', nanotechnology occupies an iconic policy space that is highly politicized. Far from an exclusively science-based initiative, nanotechnology in China has thus to be appreciated in relation to centralized 'command and control' economic planning. Nanotechnology research and development thus operates under the burdens of expected national economic transformation, the delivery of substantial commercial outcomes, the development of a knowledge-based economy, a reduction in China's technology dependence, and the flagship of China's ambitions to assume global leadership in S&T. Public perceptions of nanotechnology thus tend to be shaped in relation to sustaining and increasing national economic well being, the prospective assumption of global leadership in cutting-edge technologies and science, and improving the quality of life for Chinese citizens.

Second, the framing of nanotechnology in such overtly nationalist and aspirational contexts diminishes the political space for dissent or for the public to raise questions about safety issues, or risks associated with the impact of nanotechnology on human health and the environment. Rather, public perceptions of nanotechnology tend to be celebrated in concert with a 'rising China' and as evidence of China's destiny to assume a global leadership role. This is equally true for an increasingly large segment of the science community in China, whose livelihoods are aligned with pure and applied nano research and commercialization efforts, and who also harbor national aspirations for Chinese leadership in S&T. These dynamics portend to longer term tensions about the desire to use nanotechnology and nanoscience as flagship programs to champion China's emerging role on the world stage, while constructing regulatory modalities of governance that create effective risk management regimes to avert potential harm from nano toxicity.

Third, the command and control style approach to national economic planning and the development of nanotechnology creates elite, technocratic processes, limiting the spaces for wider consultation or public participation about the role, desirability, potential applications, and impact of nanotechnology. In a sense, the public are shut out of the policy spaces around which S&T policy is determined, or where risks and questions about potential harm from such technologies can be assessed. As observed by the head of China's NSCNN, the peak body overseeing nanotechnology in China, nanotechnology is highly technical, requires specialist knowledge, and the public does not have the technical capacities or knowledge to understand the technologies or assess potential risks (interview, 16 January 2010). Indeed, it was suggested that excluding the public or civil society groups from participation in reviews and debates was advantageous, since they might react inappropriately or form misperceptions about potential nano risks due to technical deficiencies and a poor grounding in nanoscience. Involving the public or wider non-science-based communities in discussions was thus seen as a potentially risky consultation process (see Satterfield *et al.* 2009).

Fourth, the 'knowledge deficit' problem, which in other national contexts sees the science community engage in outreach and education activities to raise evidence-based knowledge about new technologies, tends not to operate in China and the nanotechnology sector. In part this derives from a hierarchical technocratic system where there is a collusion of interests between central planners and the nanoscience community, but also in part because public perceptions toward science and scientists is deferential, with scientists highly respected and revered for their contributions to China's national economic advances. Such perceptions thus tend to reinforce the relative autonomy of the nanoscience community as the professionals most able to manage and assess the risks of nanotechnology (see Brown 2009). Such attitudes tend to moot vocal opposition, limit potential avenues for engagement between the science community and public/civil society groups, and lessen the incentives for scientists to disseminate evidence-based knowledge to the public.

Fifth, these political–social hierarchies tend to be self-reinforcing. Absent external scrutiny, public scrutiny, the ability of civil society to engage critically with evidence-based risk assessments of nanotechnologies, and concerns about the potential risks of nanotechnologies or exposure to nanomaterials are left to the science community to explore. However, the patron–client relationship that operates between central planners and the nanoscience community creates disincentives to design research programs focused on the risk impacts of nanotechnologies. Indeed, two senior Chinese nano scientists, when interviewed about possible conflicts of interest, admitted that younger scientists have incentives to under-report or downplay possible negative impacts of nanotechnologies. As they explained, funding streams for R&D are predominantly driven by the prospects for commercialization. Apart from establishing standards for nano toxicity, researchers were incentivized to open nano research avenues and not close them down through highlighting potential risks or downsides (interview, 16 January 2010).

Considerations such as these render the emergence of an effective regulatory regime able to manage nanoscience-based-risks in China problematic. The relatively closed nature of the nanoscience community, and an absence of outreach or public engagement, creates regulatory modalities that might better be characterized as self-governance or closed governance regimes. While this is not totally unusual in science-based regulation due to the technical nature of the domain, in China, where nanoscience is conflated with national economic planning, and key economic objectives focused on commercialization and international leadership, perverse incentives operate that might compromise the effectiveness of risk-based regulation. This has possible longer term implications for regulatory legitimacy in the sector, especially in the context of public risk perceptions about nanoscience and nanomaterials. Indeed, the effective side-lining of civil society engagement in the regulatory process was highlighted in numerous interviews with nano researchers who generally endorsed the need for greater civil society engagement but principally as a means to 'educate' the public so that they did not develop misperceptions about the risks associated with nanoscience and toxicity. Chinese researchers, for example, lamented the inability to communicate the benefits and risks of nano projects and ideas with the public. During interviews, they argued that Chinese society is relatively conservative and reluctant to adopt new technologies, and hinted at concerns about growing public risk perceptions of nanoscience and increasing levels of mistrust of government agencies, standards, and quality assurance regimes. Off the back of recent food scandals, contaminated baby formula, and a spate of similar quality assurance failures in domestic food manufacturing, the nanoscience community thus appears increasingly sensitized to issues of regulatory legitimacy. It remains to be seen, however, how or if this will transform the culture of regulation surrounding nanotechnology.

Appendix 9.1 List of nanotechnology-related standards in China

No.	Name	Issuer	Effective Date
GB/T13221-2004	纳米粉末粒度分布的测定X射线小角散射法 [Nanometer powder – Determination of particle size distribution – Small angle X-ray]	AQSIQ	2005-04–01
GB/T19345-2003	非晶纳米晶软磁合金带材 [Nanometer powder– Determination of particle size distribution – Small angle X-ray]	AQSIQ	2004-05-01
GB/T19346-2003	非晶纳米晶软磁合金交流磁性能测试方法 [Measuring method of magnetic properties at alternative current for amorphous and nanocrystalline soft magnetic alloys]	AQSIQ	2004-05-01
GB/T19587-2004	气体吸附BET法测定固态物质比表面积 [Determination of the specific surface area of solids by gas adsorption using the BET method]	SAC	2005-04-01
GB/T19588-2004	纳米镍粉 [Nano-nickel powder]	AQSIQ	2005-04-01
GB/T19589-2004	纳米氧化锌 [Nano-zinc oxide]	AQSIQ	2005-04-01
GB/T19590-2004	超微细碳酸钙 [Nano-calcium carbonate]	AQSIQ	2005-04-01
GB/T19591-2004	纳米二氧化钛 [Nano-titanium dioxide]	AQSIQ	2005-04-01
GB/T19619-2004	纳米材料术语 [Terminology for nano materials]	AQSIQ	2005-04-01
GB/T20307-2006	纳米级长度的扫描电镜测量方法通则 [General rules for nanometer-scale length measurement by SEM]	AQSIQ	2007-02-01
GB/T21510-2008	纳米无机材料抗菌性能检测方法 [Antimicrobial property detection methods for nano-inorganic materials]	AQSIQ	2008-08-01
GB/T21511.1-2008	纳米磷灰石/聚酰胺复合材料 第1部分: 命名 [Nano-apatite/ Polyamide composite – Part 1: Designation]	AQSIQ	2008-08-01
GB/T21511.2-2008	纳米磷灰石/聚酰胺复合材料 第2部分： 技术要求 [Nano-apatite/ Polyamide composite– Part 2: Technology requirements]	AQSIQ	2008-08-01
GB/T21511.1-2008	纳米磷灰石/聚酰胺复合材料 第1部分: 命名 [Nano-apatite/ Polyamide composite – Part 1: Designation]	AQSIQ	2008-08-01

(Continued)

No.	Name	Issuer	Effective Date
GB/Z21738-2008	维纳米材料的基本结构 高分辨透射电子显微镜检测方法 [Fundamental structures of one – dimensional nanomaterials high resolution electron microscopy characterization]	AQSIQ	2008-11-01
GB/T22458-2008	仪器化纳米压入试验方法通则 [General rules of instrumented nanoindentation test]	AQSIQ	2009-05-01
GB/T22462-2008	钢表面纳米、亚微米尺度薄膜素 深度分布的定量测定 辉光放电原子发射光谱法 [Nano Sub-micron scale film quantitative depth profile on steel –analysis – glow discharge atomic emission spectrometry]	AQSIQ	2009-06-01
GB/T22925-2009	纳米技术处理服装 [Nanotechnology-treated clothes]	China Association of Textile Industry	2009-12-01
GB/T23413-2009	纳米材料晶粒尺寸及微观应变的测定 X射线衍射线宽化法 [Determination of crystallite size and micro-strain – of nano-materials X-ray diffraction line broadening method]	AQSIQ	2009-12-01
HG/T3791-2005	氯乙烯-纳米碳酸钙原位聚合悬浮法聚氯乙烯树脂 [Suspension poly (vinyl chloride) resins via vinyl chloride and nano-calcium carbonate in situ polymerization]	NDRC	2006-01-01
HG/T3819-2006	纳米合成水滑石 [Nano-synthetic hydrotalcite]	NDRC	2007-03-01
HG/T3820-2006	纳米合成水滑石 分析方法 [Methods of test for Nano-synthetic hydrotalcite]	NDRC	2007-03-01
HG/T3821-2006	纳米氢氧化镁 [Nano-powder of magnesium hydroxide]	NDRC	2007-03-01

Notes

1 Financial support for this project was provided by the Global Public Policy Network (GPPN). The GPPN is a non-profit international partnership between Columbia University in New York, the London School of Economics and Political Science, the Lee Kuan Yew School of Public Policy at the National University of Singapore, and Sciences Po in Paris. The GPPN is aimed at fostering globally-oriented education, training, and research to address the most pressing public policy challenges of the twenty-first century. For more information about the GPPN and its activities visit www.lse.ac.uk/gppn.

2 Readers familiar with the governance of nanotechnology in China will appreciate the paucity of information available (especially in English). More obviously, the newly emerged state of the field and the rapidly changing institutional environment that oversees the sector, adds to the problems of researching the sector. We have attempted to overcome these limitations by conducting field research and undertaking a series of interviews with leading authorities in the nanotechnology arena. We are indebted to our interviewees for their time, candidness, and insights. However, the opinions expressed in this paper (as well as any possible inaccuracies) are those of the authors.

3 Carbon nanotubes (fullerenes) are derived from graphene, rolled into sheets and then tubes. They have a length to diameter ratio of up to 132,000,000 to 1, a magnitude much greater than conventional materials which endow them with unique strength, and properties such as thermal and electrical conductivity, making them ideal for incorporation into electronics and optics. See Cess Dekker (1999: 22–2). See also Appelbaum and Parker (2008:.330–1).

Bibliography

Appelbaum, Richard P. and Rachel A. Parker (2008) 'China's Bid to become a Global Nanotech Leader: Advancing Nanotechnology through State-Led programs and International Collaborations,' *Science and Public Policy*, 35(5), June: 330–1.

AOL (2010) '"The Nanotech Gamble:" Key Findings, AOL News Special Report,' 24 March 2010 <http://www.aolnews.com/nation/article/the-nanotech-gamble-aol-news-key-findings/19410735>.

Bai, Chunli (2005) 'Ascent of Nanoscience in China,' *Science Magazine*, Vol. 309, July: 61–3.

Breggin, Linda, Robert Falkner, Nico Jaspers, John Pendergrass and Read Porter (2009) *Securing the Promise of Nanotechnologies: Towards Transatlantic Regulatory Cooperation.* London: Chatham House, Royal Institute of International Affairs.

Brown, Simon (2009) 'The New Deficit Model,' *Nature Nanotechnology*, Volume 4, October: 609–11.

Chinese Academy of Science and Technology for Development (CASSTED) 'Medium to Long-term Plan for Development of Science and Technology,' (translated), Accessed 6 April 2010. <http://www.casted.org.cn/web/index.php?ChannelID=0&NewsID=3747>.

Dekker, Cess (1999) 'Carbon Nanotubes as Molecular Quantum Wires,' *Physics Today* 52: 22–2.

Drexler, K. Eric (2006) *Engines of Creation 2.0: The Coming Era of Nanotechnology.* Twentieth Anniversary Edition. Available at http://e-drexler.com/p/06/00/EOC_Cover.html.

Falkner, Robert (2008) 'Who's Afraid of Nanotech?' <Theworldtoday.org> June: 15–7.

Lerwen, Liu and Li De Zhang (2005) 'Nanotechnology in China _ Now and in the Future,' *Nanotechnology Law & Business*, November/December: 397–403.

Lindquist, Eric, Katrina N. Mosher-Howe and Xinsheng Lui (2010) 'Nanotechnology....
What is it Good For? (Absolutely Everything): A Problem Definition Approach,' *Review of Policy Research*, 27(3): 255–71.

Li, Liu and Jingjing Zhang (2007) 'Characterising Nanotechnology Research in China', *Science Technology Society*, 12: 201–16.

Lux Research <http://www.luxresearchinc.com/index.php>

Ministry of Science and Technology of China (MOST) 'National High Tech R&D Program (863 Program)', Accessed 6 April 2010 <http://www.most.gov.cn/eng/programmes1/>.

Renn, O and M. C. Roco (2006) 'Nanotechnology and the Need for Risk Governance,' *Journal of Nanoparticle Research*, 8: 153–91.

Satterfield, Terre, Milind Kandlikar, Christian E. H. Beaudrie and Joseph Conti (2009) 'Anticipating the Perceived Risk of Nanotechnologies,' *Nature Nanotechnology*, 265, 20 September: 1–7.

Scheufele, Dietram A., Elizabeth A Corley, Sharon Dunwoddy, Tsung-Jen Shih, Elliott Hillback and David Guston (2007) 'Scientists Worry about Some Risks more than the Public ,' *Nature Nanotechnology*, Volume 2, December: 732–4.

Shapira, Philip and Jue Wang (2009) 'From lab to Market? Strategies and Issues in the Commercialization of Nanotechnology in China,' *Asian Business and Management*, 8(4): 461–89.

Scientific American (May 2008) 'Study says Carbon Nanotubes as Dangerous as Asbestos,', Accessed 21 September 2010 <http://www.scientificamerican.com/article.cfm?id=carbon-nanotube-danger>.

Woodrow Wilson International Center and the Pew Charitable Trusts (2010) *Project on Emerging Nanotechnologies*, <http://www.nanotechproject.org/>.

World Bank 'World Development Indicators.' Accessed 6 April 2010, <http:// www.worldbank.org› Data>.

10 Lessons from biomedical technology regulation

North American and European comparisons[1]

Isabelle Engeli, Christine Rothmayr Allison and Frédéric Varone

Introduction

Spurred by the general advancement of biomedical research, the spectrum of available assisted reproductive technologies (ART) and embryo related research has considerably broadened during the last three decades. The first *in vitro* fertilisation (IVF) birth in 1978 in the UK has opened the door for a number of further breakthroughs. The development of preimplantation genetic diagnosis (PGD) in 1990 soon allowed couples with severe hereditary illnesses to select embryos that were not carriers of the defective gene (see Handyside, Kontogianni, Hardy, and Winston 1990) and the first intracytoplasmic sperm injection (ICSI) performed by a Belgian team in 1992 has helped overcome severe male infertility without requiring donor sperm. Research on leftover embryos contributed to developing embryonic stem cell research, with the first derivation of human embryonic stem cells (ESCR) in 1998, leading up to the development of therapeutic cloning at the beginning of the new millennium. Technological advances and the wide commercialisation of ART treatments created controversies around the globe (Gaskell and Bauer 2001). Alleged attempts to clone humans, various scandals linked to clinical practice, high medical profiles for post-menopausal mothers, and legal battles regarding the funding of stem cell research, have contributed to ART's salience in media and politics.

In Western democracies, policymakers have addressed the challenges of ART in contrasting ways (e.g. Fink 2008; Montpetit *et al.* 2007; Bleiklie *et al.* 2004; Engeli 2009). Some countries regulated ART early on, mostly addressing issues related to IVF, at a time when PGD and ESCR were still more hypothetical than real. They then adapted their first generation policies to the second generation of ART and embryo related research through revising existing laws or adopting additional legislation. Other countries addressed ART much later after the breakthrough in ESCR in 1998.

Our contribution analyses, first, whether and why early and late movers have adopted fundamentally different policies. Our second research question pertains to the development of policies over time: how and why have countries moved from first to second generation regulations, and which factors account for policy stability or change over time? We are particularly interested in how first generation regulation has impacted second generation policies, and also ask whether second generation regulations are more similar across countries than first generation policies. Specifically, this chapter analyses policies addressing assisted reproductive technology and embryonic stem cell research for nine European and North American countries: the USA, Canada, Belgium, France, Germany, Italy, Netherlands, Switzerland, and the United Kingdom.

Theorising policy trajectories: path dependency, policy transfer, and scientific development

Existing literature points to a number of competing explanations for policy development over time. However, there has been no systematic testing of these hypotheses for a broad range of countries and assisted reproductive techniques over time. An early prominent explanation refers to the importance of existing regulatory regimes for addressing new challenges in green and red biotechnology alike. For ART treatments and research, path dependency (Pierson 2000, 2006) or policy heritage (Rose and Davies 1993) approaches are of particular relevance because earlier breakthroughs, such as the development of IVF, were necessary conditions for later breakthroughs, such as PGD, ESCR or therapeutic cloning. Accordingly, early regulations addressing such issues as IVF and related research – especially if those restricted the existing practice – also spoke to or even covered later developed techniques. Hence, more 'advanced' techniques can easily be assimilated to earlier debates and policies. Nevertheless, we are not arguing that technological development per se explains policy stability over time. Rather we suggest that path dependency can be explained in the following way. First, we argue that we can observe lock-in effects related to the considerable costs of debating 'moral' policy issues, which strongly polarise political actors, and also potentially divide political parties and governmental coalitions. Thus, sticking to an already established consensus or decision reduces costs associated with reopening the debate and also the risk of important conflicts within political parties or governmental coalitions (Banchoff 2005). Furthermore, policies already in place have an impact on actor constellations; they give a comparative advantage to those defending the status quo in comparison to forces seeking fundamental policy change (Pierson 2000). If during a first generation of regulation, for example, medical and research interests have established privileged networks with key administrative political actors resulting in policies favouring instead of limiting ART, these policies strengthen the already established privileged relationship and give research and medical interests an important say in further developing ART related policies. Accordingly, the first hypothesis to test is (H1): Early regulators of ART and embryo related research[2] continue in their general policy path. That is, according to H1, countries that were initially liberal remain liberal, and countries that were initially restrictive remain restrictive.

However, the literature also argues that there are a number of factors that might induce policy change towards more uniform, and especially more permissive policies over time. The first perspective is based on an economic, interest-driven account of policy development. Incentives for minimal or permissive regulations, possibly combined with strong promotional activities, increased with the number of scientific breakthroughs in the late 1990s. International research competition and the future economic and public health benefits of stem cell-based therapies set a strong incentive for choosing policies that increase instead of decrease national competitiveness. A second reason for expecting countries to adopt more permissive or less restrictive policies from the 1990s onward might be processes of policy transfer. Policies already in place in other countries play an important role in biotechnology policy making as issues become politically salient at different points in time. Some countries are leaders in research and commercialisation, while others are late adopters from a technological point of view. Furthermore, in some countries controversies remain very limited, putting experts at the centre of policy formulation. When permissive regulations in other countries do not lead to a wave of scandals and negative developments, it tends to undermine the slippery slope argument that allowing some research or practice will open the door to more unwanted developments. The evolution of public opinion might be a third reason for the likelihood of permissive policies increasing over time. Public opinion research reveals considerable differences in attitudes towards biotechnology across countries but also according to the type of application, with biomedical applications more often favourably evaluated than agro-food biotechnology (Bauer 2005; Gaskell and Bauer 2001). ESC research as a means for finding new treatments has certainly opened the door to more positive framing of ART and embryo related research. This more positive framing of ESC research in terms of the potential contribution to improving our health likely contributes to greater acceptance of ART and research in general. In addition, science and technology studies also point out that new technologies get better accepted over time and this is also likely the case with ART. Hence, the second competing hypothesis is (H2): Western countries converge towards intermediary and/or permissive policies from the end of the 1990s onward. Late regulators tend towards intermediary and permissive designs while early regulators revise their policies in a more permissive way.

Comparing regulatory trajectories for ART and embryo related research

This section analyses and compares patterns in regulating ART and embryo related research. We conceptualise medical and scientific autonomy as the degree of freedom granted to doctors and researchers to decide upon both the technology to be used and the conditions under which they should be used (Bleiklie *et al.* 2004). To capture policy change over time, the level of medical autonomy is measured at each policy decision time point for five different biomedical applications: *in vitro* fertilisation (IVF), preimplantation genetic diagnosis (PGD), embryonic stem cell research (ESCR), reproductive cloning (RC), and therapeutic cloning (TC).[3] In order to render our analysis more readable, for the remainder of the text we will simply distinguish between ART and embryo related research, where the first covers IVF and PGD, and the second includes ESCR, RC, and TC.

Biotechnologies are *permissively* regulated (score 3) if they are not subject to any substantial restriction (that is, are generally permitted) but might be conditioned by licensing/reporting procedures. The regulation is qualified as *intermediate* (score 2) if some light restrictions are imposed. Biotechnologies are *restrictively* regulated (score 1) if their use is severely constrained. Finally, they could be outright *banned* (score 0). In order to speak of major changes, we would expect a country to shift across two categories – e.g. from banned (0) to intermediate (2), or from restrictive (1) to permissive (3) – for a broad spectrum of techniques. Incremental change would consist of minor shifts regarding some issues only – e.g. PGD moves from intermediate (score 2) to restrictive (score 1) – while policies for the other issues remain stable.

Figures 10.1 and 10.2 plot regulatory trajectories for ART and embryo related research from 1980 to 2010 for seven West European (Italy, Belgium, Switzerland, Germany, the United Kingdom, the Netherlands, France) and two North American (Canada and the United States) countries. At first glance, the results clearly contradict the overall convergence hypothesis (H2). Increasing economic competition over human biotechnology between countries and continuing scientific breakthroughs have not resulted, so far, in broad convergence toward permissive regulation. On the contrary, confronted with rapid and cutting edge developments in human biotechnology, governments have adopted strongly diverging policies, ranging

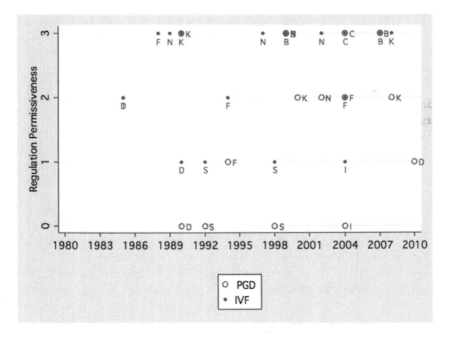

Figure 10.1 Patterns in regulatory trajectory over ART

Note: 'B': Belgium; 'C': Canada 'D': Germany; 'F': France; 'I': Italy; 'K': The United Kingdom; 'N': The Netherlands; 'S': Switzerland; 'U': The United States

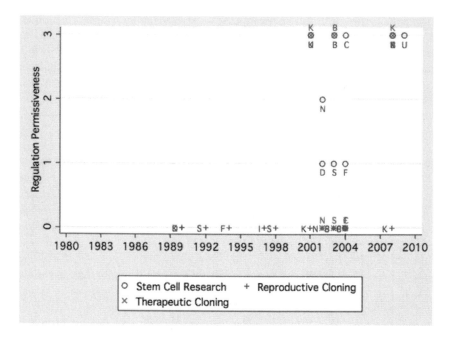

Figure 10.2 Patterns in regulatory trajectory over embryo related research

Note: 'B': Belgium; 'C': Canada 'D': Germany; 'F': France; 'I': Italy; 'K': The United Kingdom; 'N': The Netherlands; 'S': Switzerland; 'U': The United States

from fully prohibiting some technologies to broad permissiveness for the entire sector (Bleiklie *et al*. 2004; Montpetit *et al*. 2007; Engeli 2009). The one exception is RC that is now banned by all the countries in our sample except the USA.

The time of adoption also varies considerably across countries and issues. We can distinguish three clusters of countries. In a *first* cluster, the 'early mover' countries (Germany, Switzerland, the Netherlands, and the UK) responded quickly to scientific advances in the 1980s and in the beginning of the 1990s and their initial policies exerted a strong path dependent effect on their way of dealing with major breakthroughs in ESCR and TC from the late 1990s onward. Germany and Switzerland have remained restrictive, while the Netherlands and the UK have not deviated from their initial permissive path. Belgium, Canada, and the US belong to another, *second* cluster of 'late mover' countries that refrained from substantially regulating ART and embryo related research until the 2000s. Their regulations then followed a permissive path. Finally, France and Italy form a *third* cluster because they strongly deviated from their initial regulatory path and became more restrictive over time.

In the 'early mover' countries, initial policies primarily addressed IVF and were enacted in response to the rapid diffusion of ART. In 1985, Germany was first to address ART and was quickly followed by the Netherlands in 1989, then the

United Kingdom in 1990 and Switzerland in 1992. Their early launch of regulation processes has not resulted, so far, in any convergent effect towards permissiveness. On the contrary, Germany and Switzerland opted for an overall restrictive regulation that imposed severe restrictions on ART, while the Netherlands and the UK chose to govern this new medical field more permissively. Initial regulation on ART has exerted a considerable influence on how the early mover countries later dealt with technological breakthroughs in ESCR and TC. In Germany, Switzerland, the Netherlands, and the UK, policies evolved at their own national pace. On the one hand, the Netherlands and the UK followed a permissive path. The 1990 UK Human Fertilisation and Embryology Act provided a high level of autonomy to practice ART and conduct embryo research and was extended in 2001 to stem cell research and TC. In the same vein, the Netherlands first permissively regulated ART and then designed a research friendly policy regarding stem cell research in the 2002 Embryo Act. On the other hand, Germany and Switzerland have not deviated from their initial restrictiveness. The severe restrictions they imposed on ART practices were extended to stem cell research, all forms of cloning were banned, Germany allowed PGD only under severe restrictions in 2010, and Switzerland has (so far) maintained the ban.

Two 'early mover' countries have strongly deviated from their initial path over time and thus form another cluster: France and, in a more radical way, Italy. France initially proposed a procedural response in 1998 to ART development by introducing a licencing procedure for the ART centres. In 1994, The Bioethics Laws imposed some light restrictions on ART and, above all, severely constrained the use of PGD. In 2004, the revision of The Bioethics Laws confirmed the move toward more restrictiveness: in a similar fashion to Germany and Switzerland, stem cell research has only been allowed under severe restrictions and TC has been banned. Italy has gone even further than France. Their initial regulations in 1985 and 1987 were rather permissive and imposed only light restrictions on ART, mainly in the public health sector. However, in 2004, a dramatic policy change occurred. ART was severely restricted and a total ban was imposed on PGD, ESCR, and TC all together.

Three countries can be best qualified as 'late movers' – as mentioned, Belgium, Canada, and the USA. No regulation was initially imposed and when they ultimately were, they were permissive. Nevertheless, they strongly vary in terms of policy scope among each other. The USA, at the federal level, has only addressed issues regarding the public funding of embryo related research, while Belgium and Canada have designed more comprehensive regulation. In a similar fashion to France, Belgium's initial regulation in 1999 only imposed licensing obligations on ART centres. Contrary to France, subsequent Belgian regulation remained permissive and provided the medical and scientific communities with significant autonomy in developing stem cell research and therapeutic cloning. Canada did not have any comprehensive regulation of ART and embryo related research until 2004. The Assisted Human Reproduction Act grants physicians and scientists considerable autonomy regarding ART, PGD, and ESCR but bans reproductive and therapeutic cloning.

Case studies: explaining trajectories

This section looks at the three clusters of countries in more detail to understand better which factors explain their policy choices and regulatory trajectories over time.

Early mover cluster

Except for RC, the two sets of countries in this cluster addressed the initial challenges of IVF and the later breakthroughs in ESCR very differently. As described above, Switzerland and Germany, have adopted very restrictive policies, while in the United Kingdom and the Netherlands the regulations are permissive. Yet, both sets of countries depict strong patterns of path dependency over time: the first two continuing on their initially restrictive path, while the latter two stuck to their much more permissive trajectory.

Restrictive early movers

The German and Swiss cases are both characterised by an early and broad mobilisation against ART and biotechnology more generally (see Rothmayr and Serdült 2004; Rothmayr and Ramjoué 2004; Rothmayr 2006; Abels and Rothmayr 2007; Engeli 2009, 2010). In neither case did the debate focus on whether to adopt intermediary or restrictive solutions, but on whether it would be preferable to prohibit IVF outright.

Past abortion debates, and the experience of World War II, are important for understanding the German Embryo Protection law (EschG), adopted in 1992 under a coalition government of Christian Democrats (CDU/CSU) and Liberals (FDP). Based on the 1975 decision of the Federal Supreme Court to strike down the abortion law, the Christian Democrats advocated for strong protection of the embryo. The Christian Democrats faced opposition from the left. The left feared that ART would open the door for eugenic uses and therefore asked to prohibit IVF entirely. Furthermore, mobilisation against ART was also motivated by distrust of the scientific and political elite that characterised the strong German social movements of the 1980s. Against this constellation of actors seeking very restrictive policies, the German Research Council, together with other research interests, did not succeed in defending their position that embryo research should be permitted under certain conditions based on the constitutionally guaranteed freedom of science. They were somewhat more successful with ESCR in the late 1990s because they had a strong ally in Chancellor Schröder. The derivation of stem cells in Germany would have demanded a change of the existing legal framework, but the governing Social Democrats and Greens were divided over the issue, and so were the other parties. Hence, Parliament struck a compromise between the advocates of stem cell research and their opponents. Parliament did not revise the EschG, but allowed for the *import* of stem cell lines from abroad under specific conditions. While the adoption of the Stem Cell Law in 1998 temporarily solved the research question, other issues remained on the political agenda. In 2010, the

German Federal Court decided that the use of PGD to prevent the transmission of grave hereditary diseases is not a criminal offence, thus opening the door for further political debate and possible policy changes.

In Switzerland, direct democracy provided important venues for influencing policy outcomes. The first initiative, the 'Beobachter' initiative launched by consumer protection groups in the mid-1980s, succeeded in framing the issue of ART in the federal constitution as a potentially dangerous technology needing state intervention in order to protect people from potential misuse. There was considerable support for total prohibition from both sides of the political spectrum: Christian religious beliefs in the Centre and the Right, and a sceptical attitude towards scientific progress on the Left. Parliament formulated a restrictive counter-proposal in order to counteract the demand for a total ban of IVF and related techniques. The counter-proposal satisfied the sponsors, who withdrew the initiative, and was accepted by the people in a popular vote in 1992. The second popular initiative, the 'Initiative for Procreation respecting Human Dignity' sought to reverse the constitutional article by prohibiting IVF and insemination by donor in the constitution. The initiative did not pass the popular vote (in 2000) but was a 'pledge' in negotiating policy solutions in the pre-parliamentary and parliamentary stages (Linder 1994; Papadopoulos 2001) of elaborating the federal law on ART. The Federal Council (government) proposed a restrictive federal law on ART in order to counteract the second popular initiative. Again, against a Left–Centre–Right coalition, the Liberal Party together with medical interests advocated for a more permissive solution, yet without success. Shortly after, however, research and medical interests were successful in realising their common goal to allow ESCR in Switzerland in the Stem Cell Research Act because they mobilised more strongly and because of a more favourable actor constellation. There was comparably less mobilisation against ESCR. Moreover, they had a strong partner in the Office of Public Health, who was in charge of elaborating the draft and had defended throughout the process a more research friendly approach, in comparison to the Justice Department, which lead the earlier legislation. The optional referendum against the Stem Cell Law was launched by anti-abortion and anti-biotech interest groups, yet the law was accepted in the popular referendum with a clear majority.

Permissive early movers

While critical concerns over scientific abuses and social implications of ART were rapidly raised in Germany and Switzerland, ART developments in the Netherlands and in the UK have received more favourable support overall within public opinion. In both countries, ART related issues were, in the first stage, depoliticised and delegated to expert commissions and specialised regulatory bodies. It is only in a second stage that these issues were substantively addressed by the political agenda through design of comprehensive regulation. The comprehensive regulation has not dramatically departed from the initial permissive regulation and the medical and scientific communities have been granted broad autonomy.

In the Netherlands, IVF as a treatment of infertility has benefitted from strong public support since the beginning of the 1980s (Timmermans and Scholten 2006). The number of ART centres increased rapidly, and the medical community set up self-regulating guidelines that were well respected by the practitioners (Timmermans 2007). Political parties were reluctant to put on the agenda any issue that could re-launch the controversy over abortion (Timmermans 2007). As a result, the ART issue quickly depoliticised and was delegated to science-related venues and specialised advisory bodies (Timmermans and Scholten 2006). During the 1980s and 1990s, the government refrained from any substantive intervention in the field of ART. Instead, it only issued procedural decrees, mainly by imposing ART centres licensing procedures in 1989, 1998, and 2000, with the broad support of the medical community. Nevertheless, breakthroughs in embryo related research during the 1990s put the issue back on the political agenda. The exceptional absence of the Christian Democrats from the government coalition between 1994 and 2002 allowed the formation of a secular coalition that was ready to address how best to govern the ART field in a substantive way (Timmermans 2004). The resulting 2002 Embryo Act is one of the most permissive regulations in Europe regarding embryo related research. While it imposes a temporary ban on embryo creation for research purposes and therapeutic cloning, the Act provides the science community with a lot of autonomy in conducting research on ESC derived from spare IVF embryos. The reintegration of Christian Democrats in the government coalition in 2002 has not resulted in any major revision of prior regulations, but has rendered unlikely a lift on the temporary ban on embryo creation for research purpose in the near future (Timmermans 2007).

The United Kingdom became a leading country in the ART field with the world's first successful IVF in 1978. The government quickly considered the ART issue to be a 'politically hot potato' and mandated a Committee of Inquiry to make recommendations regarding the social and ethical implications of ART technological developments (Blank 2004: 123). Meanwhile, medical associations were also willing to take the lead on ART issues and produced several reports in favour of medical self-regulation. The polarisation of the public debate increased when a private conservative bill on the 'protection of the unborn children' received strong support in Parliament (Blank 2004). Facing upcoming parliamentary elections, the government preferred to depoliticise the ART issue one more time and submitted a regulatory framework for consultation (Blank 2004). In the meantime, the medical community set up a voluntary licensing authority to prove their willingness to establish a set of good ART practices (Blank 2004). The Human Fertilisation and Embryology Act approved in Parliament in 1990 largely corresponded to the policy preferences of the medical community in favour of broad autonomy (Montpetit 2007). Indeed, the Act established a regulatory agency in charge of licensing and controlling ART centres, elaborating medical guidelines, and allowing embryo related research projects. The only banned technique was RC, which had the approbation of the medical community. The regulatory agency provided the medical and science communities with direct access to the policy making process and enabled them to maintain considerable influence over time (Montpetit 2007). In 2001, the Act was revised in order to allow for ESCR, including therapeutic cloning and the creation of embryo for research purposes, while the Human Reproductive Cloning Act was issued in the same year to reinforce the ban over reproductive cloning. In 2008, under the active promotion of stem cell research by the Blair government, the revision of the Human Fertilisation and Embryology Act confirmed the permissive path adopted by the United Kingdom in the field of ART and embryo related research (Larsen *et al.* 2012).

Policy change cluster

While the regulatory trajectories in Germany, Switzerland, the Netherlands, and the UK have been strongly influenced by the initial regulations, France and Italy have considerably deviated from their initial regulatory path and have become more restrictive over time. In both countries, electoral shifts from Left to Right wing governments, combined with a continuously strong Catholic opposition to ART, proved to be decisive for the radical change in ART and embryo related research policies.

In France, the first phase of the policy making process began with the increased availability of IVF in the early 1980s, which led to a steamy public debate on the purpose of science and the social implications of ART development. Numerous ART centres opened rapidly, and competed with each other to be at the cutting edge of ART technological development, while major medical associations were, on the contrary, reluctant to approve ART techniques and did not take the lead on developing medical self-regulation. Unable to rely on any consensus within the medical community about how to best regulate the field, the government quickly postponed the elaboration of a comprehensive law and, instead, merely issued a procedural decree requiring the licensing of ART centres in 1988 (Engeli 2004). During the early 1990s, the development of PGD incited the Left wing government to resume its efforts to substantively govern ART. Specialised ART practitioners were provided with ample room in the debate and were recognised as having a status at least on par with that of the representatives of the main medical associations. As a result, the draft bill, submitted in 1992, was rather permissive and granted the medical community considerable autonomy. The right wing coalition that came to power in 1993 re-designed the draft bill and added some substantive restrictions on practicing IVF and PGD as well as dramatically restraining embryo related research. With the support of the Catholic–conservative members of the coalition, the laws on bioethics were adopted in 1994. In the beginning of the 2000s, the ESCR breakthrough led the left wing government to revise the laws on bioethics in order to render them more permissive regarding embryo related research. Nevertheless, the 1993 parliamentary elections repeated the 1994 scenario. A right wing majority came to power again and, again, revised the law on bioethics to be more restrictive. Consequently, the 2004 law bans TC and allows for ESCR only under very restrictive conditions (Engeli 2010).

In Italy, the first phase of the policy process resembles the French case. In response to the first IVF birth, attempts to address ART in Italy began in 1983. For the next two decades, however, regulations remained limited. Two circulars issued by the Ministry of Health in 1985 and 1987 constituted the core regulation of ART. As a reaction to the AIDS crisis, these circulars were aimed at preventing the transmission of HIV through sperm donation. Until 2004, Italian policy was thus among the most permissive in Europe and North America, commonly described as the lawless 'Far West' of fertility treatments (Ramjoué and Klöti 2004). The new centre–right dominated legislature, elected in May 2001, eventually adopted, in February 2004, a first law entitled 'Norms regarding medically assisted reproduction'. This law, still in force today, represents a radical change and is considered one of the most restrictive regulations on ART currently in place. It bans ESCR, PGD, and IVF with gamete donation. The leftist libertarian and anti-Catholic Radical Party successfully launched an abrogative referendum against this law, with the support of various women's and patients' interest groups. On the opposite side, the Roman Catholic Church successfully urged Italian citizens to abstain from voting. The popular vote organised in June 2005 indeed yielded only a twenty-six per cent turnout. Since constitutional norms in Italy require a minimal turnout of fifty per cent for the results of a referendum to be valid, the outcome of this popular vote (i.e. a revision of the law accepted by eighty-eight per cent of the voters) was declared void. This spectacular U-turn of Italian policy represents an exceptional 'policy punctuation' from one of the most permissive to one of the most restrictive regulations in Europe and North America, and can best be explained by three factors. First of all, the Centre–Right parties forming the governmental coalition all shared the Catholic Church's opposition to ART. Furthermore, the Catholic Church also invested considerable resources in policy advocacy and referendum campaigning. Finally, in sharp contrast to the religiously motivated activism, the proponents of more permissive regulations had diverging interests and beliefs regarding embryo related research and did not mobilise enough (Schiffino *et al.* 2009).

Late mover cluster

As mentioned, the late mover cluster is characterised by three countries: Belgium, Canada, and the USA. All three currently have very permissive policies. Yet, the reasons for their rather late adoption of policies vary considerably across the three countries.

The United States is mostly characterised by the absence of binding decisions on the federal level. In contrast to Canada and Belgium, in the USA all attempts to address ART comprehensively on the federal level have failed.[4] In fact, policy debates and policies in the USA have mainly focused on the question of embryo and stem cell research (Bonnicksen 2002; Goggin and Orth 2004; Garon and Montpetit 2007). Yet even for embryo related issues, policies have remained very limited.[5] The conflict around ESCR is structured along the fault lines of the abortion debate, with pro-life and pro-choice groups taking sides for or against embryo related research. Furthermore, patient groups and research interests have mobilised in order to support ESCR in light of its potential future benefits in curing various degenerative and other diseases, for example Parkinson's disease. The association with the abortion debate made ESCR a highly salient electoral issue, which is a unique situation among the countries in our sample. The strong divisions along the lines of the abortion debate, together with the numerous institutional veto-points in the US political system, rendered any *comprehensive* regulation of ESCR impossible to achieve. In 1995, Congress – under a new Republican majority – passed the Dickey-Wicker Amendment to the National Institutes of Health Revitalization Act that prohibited only federal public funding for any embryo research destroying or harming an embryo *in vitro*. In 2001, President Bush reacted to the breakthroughs in ESCR by permitting the financing of research on stem cell lines originating from leftover embryos (with the consent of the couple) that were created prior to his decision. In 2010, by executive order, President Obama allowed public funding to be used on new stem cell lines. Yet, this decision has been attacked in court on the grounds that the Dickey-Wicker Amendment is still in force. The court proceedings have not yet concluded. However, the impact of federal policies on medical and research autonomy remains limited, as they do not impose any limitations or substantial regulations on *privately* funded ESCR research or research funded by individual states, as is, for example, the case in California.

Like the USA, Belgium has been a pioneer in developing ART and embryo related research and has shown pronounced reluctance to intervene in the biomedical sector. Frequent attempts to regulate ART in a substantive and comprehensive manner failed in the 1980s and 1990s. Procedural arrangements, such as licensing schemes for ART centres (in place since 1999) were adopted instead. Two main factors explain this 'legal void'. First, Belgian governments were dominated by Christian Democratic parties until 1999. In order to prevent potential divisions on ART policy within governing coalitions, and to secure their own internal party cohesion as well, Christian Democrats deliberately kept all morality issues off the political agenda. On the other hand, physicians favour self-regulation that grants them a high degree of autonomy. Officially, the medical profession was in favour of licensing arrangements in order to guarantee the quality of ART procedures. Increased economic competition from a growing number of ART centres, however, also motivated the medical profession to support licensing measures (Varone and Schiffino 2004). The regulatory situation has changed dramatically during the present decade, but from within the same permissive framework. In May 2003, Belgium passed a law on research on *in vitro* embryos, which authorises the procuring of stem cells from residual embryos, therapeutic cloning, and the creation of embryos for research purposes. Only RC is explicitly forbidden. Moreover, in July 2007, Belgium adopted another law authorising ART for singles and homosexuals, PGD, and post-mortem insemination. The content of the policy design is thus still very permissive and shows a clear path dependency from the previous procedural rules in use. However, Belgium now has a formal and substantive regulation of ART and embryo related research.

This change was made possible by the formation of a new secular governing coalition in 1999 consisting of Socialists, Liberals, and Greens, and excluding the Christian Democrats. This trend was maintained by the subsequent secular coalition of Socialists and Liberals, not including the Greens and Christian Democrats, from 2003 until 2007. Not surprisingly, physicians were in favour of this new policy design that was obviously not limiting their autonomy, while opponents to ART and embryo research (i.e. religious groups) did not mobilise at all – in contrast to the situation in Italy (Schiffino *et al.* 2009).

Canada's road to a comprehensive regulation has been equally long. The Assisted Human Reproduction Act (AHRA) criminalises the remuneration of surrogate mothers and the purchase and the sale of reproductive and other human material, prohibits the creation of embryos for research purposes, bans all forms of cloning (including therapeutic cloning), but allows for ESCR on left over embryos. The Act was adopted following the tenacious work of consultation and formulation that lasted more than ten years (for details see Montpetit 2004; Montpetit 2007; Scala 2003) beginning in the early 1990s. The Liberal Government under Chrétien proposed a bill in 1996, yet this bill was never adopted because of several overlapping lines of conflict (Montpetit 2004; 2007). The medical profession was opposed to criminal prohibitions and considered self-regulation within the provincial health care systems sufficient. A number of provinces were opposed to federal regulations, because they considered them to intervene into their health care prerogative, a view partly confirmed by a recent Supreme Court decision.[6] The women's network wanted to see regulations, but was opposed to any criminal law intervening into women's reproductive freedom.

In fact, the legacy of the abortion debate considerably contributed to structuring policy making processes. Networks involved "…rested on the relationship that members of Parliament (…) had developed with physicians in their constituency during the abortion debate of the early 1990s" (Montpetit 2007: 92) giving a predominant place to physicians and women in the debate at the expense of right wing groups advocating for more restrictive regulations.[7] The bill also failed because of electoral considerations (Montpetit 2007: 96): "given that the issue of embryo related research divided the Liberal Parties supporters, and in order to avoid electoral costs in the wake of an upcoming election, the government decided to let the bill die on Parliament's order". In 2004, a law on ART was finally adopted. Health Canada has considerably changed its consultation strategy, facilitating broader support by being more inclusive (Montpetit 2007). At the same time, the women's network developed a common focus on ART as primarily a health issue, rather than an issue regarding reproductive rights, and one that should be regulated to protect women's security and well-being (Montpetit 2007). Furthermore, breakthroughs in ESCR rendered a regulatory framework even more pressing. Finally, changes in the party landscape decreased the Conservative and Liberals interest in turning ART and embryo related research into an electoral topic because of the risk of losing voters, leaving the regulation largely to the network dominated by medical and health interests (Montpetit 2007).

Conclusion

Our empirical comparison of the policies regulating ART and embryo related research in nine European and North American democracies demonstrates that countries do not converge over time towards more permissive policies granting greater autonomy to physicians and researchers. In other words, economic, scientific, or public health interests in biomedical breakthroughs do not induce a regulatory race-to-the-bottom. On the one hand, the empirical evidence shows that late movers (i.e. Belgium, Canada, and the USA) adopt permissive policies. On the other hand, regulatory patterns over time indicate that early movers, who initially regulated ART in a restrictive manner (i.e. Germany and Switzerland), do not follow a trajectory with major policy punctuations towards more liberal policy designs. However, we observed the reverse situation in Italy and France, where both countries experienced a major policy change translating into less permissiveness – and not into more permissiveness as expected by our theoretical framework. Last but not least, the large majority of our cases are best characterised by a path dependent pattern, with only some incremental policy changes occurring during the last three decades.

Three lessons can be drawn regarding the main factors explaining regulatory trajectories of biomedical policies in Western democracies. First, party politics matter, even in the USA (i.e. electoral campaign). Advocacy coalitions built upon Christian Democrats and the biotech-adverse New Left strongly influenced the adoption of restrictive ART policies, while permissive policies were implemented by secular parties (Engeli *et al.* 2012). Second, interest groups matter as well, as illustrated by the policy impacts of physicians and religious actors' mobilisation and framing strategies (Montpetit *et al.* 2005; Engeli 2012). Physicians' self-regulation, as well as the recommendations formulated by expert groups in the very first stage of the decision-making process, impact on the resulting policy (Rothmayr 2003) and, furthermore, tend to depoliticise the 'morality issues' at stake. On the opposite side, the political activism of religious interest groups significantly increases the degree of conflict and the potential for restrictive ART and embryo related research policies. Third, our investigation failed to identify any causal link between the types of political systems (centralised versus federalist country, majoritarian versus consensus democracy, etc.) and the regulatory trajectory of ART and embryo related research (Rothmayr *et al.* 2004). This does not mean that institutions do not matter. The frequent use of direct democratic devices in Switzerland and Italy provides enough evidence about the important role of institutions. Rather, it means that an actor-centred institutionalism approach is adequate to analyse the joint effects of actor interests, institutional venue shopping, and problem framing.

Notes

1 Isabelle Engeli and Frédéric Varone acknowledge the financial support of the Swiss National Science Foundation (project ref. 105511-119245/1).
2 That is, countries which have elaborated a first policy before the mid-1990s.
3 While concentrating exclusively on the autonomy granted to the medical and science communities in practicing ART and conducting embryo related research, this chapter does not cover the access granted to patients to ART. It also does not cover related techniques such as surrogacy and specific regulations regarding gametes and embryo cryopreservation as well as donor's consent.
4 Legislation at the state level has also remained limited in scope and nature.
5 Only reproductive cloning is prohibited in California, Connecticut, Maryland, Massachusetts, New Jersey, and Rhode Island. Both types of cloning are prohibited in Indiana, Iowa, Michigan, North Dakota, South Dakota, and Virginia. There are prohibitions of state funding for reproductive cloning in Missouri and for reproductive and therapeutic cloning in Arizona.
6 The Supreme Court decision invalidated a number of provisions regarding ART, but has no impact on the overall classification of the Canadian law as permissive.
7 As Francesca Scala argues, for ESCR and embryo research in general, feminists took a critical stance towards the scientific and medical discourse, which dominated political deliberations (Scala 2003: 86).

Bibliography

Abels, Gabriele and Christine Rothmayr (2007) 'ART and GMO Policies in Germany: Effects of Mobilization, Issue-Coupling, and Europeanization', in Éric Montpetit, Christine Rothmayr and Frédéric Varone (eds) *The Politics of Biotechnology in North America and Europe*. Lanham: Lexington (145–68).

Banchoff, T. (2005) 'Path Dependence and Value-driven Issues. The Comparative Politics of Stem Cell Research', *World Politics*, 57: 200–30.

Bauer, Martin W. (2005) 'Distinguishing Red and Green Biotechnology: Cultivation Effects of the Elite Press', *International Journal of Public Opinion Research* 17(1): 63–89.

Blank, Robert H. (2004) 'The United Kingdom: Regulating Through a National Licensing Authority', in Ivar Bleiklie, Malcolm L. Goggin and Christine Rothmayr (eds) *Comparative Biomedical Policy. Governing Assisted Reproductive Technologies*, London: Routledge (120–37).

Bleiklie, I., M. L. Goggin and C. Rothmayr (eds) (2004) *Comparative Biomedical Policy*. London: Routledge.

Bonnicksen, Andrea L. (2002) *Crafting a Cloning Policy: from Dolly to Stem Cells*. Washington, DC: Georgetown University Press.

Engeli, I. (2004) 'France: Protecting Human Dignity while Encouraging Scientific Progress', in I. Bleiklie, M. Goggin and C. Rothmayr (eds) *Comparative Biomedical Policy: Governing Assisted Reproductive Technologies*. London: Routledge.

Engeli, I. (2009) 'The Challenges of Abortion and Assisted Reproductive Technologies Policies in Europe', *Comparative European Politics* 7(1): 56–74.

Engeli, I. (2010) *Les Politiques de la Reproduction*. Paris: L'Harmattan.

Engeli, I. (2012) 'Policy Struggle on Reproduction: Doctors, Women, and Christians', *Political Research Quarterly* (forthcoming).

Engeli, I., C. Green-Pedersen and L. Thorup Larsen (2012) *Morality Politics in Western Europe. Parties, Agenda and Policy Choices.* London: Palgrave Macmillan.

Fink, S. (2008) 'Politics as Usual or Bringing Religion Back In?: The Influence of Parties, Institutions, Economic Interests, and Religion on Embryo Research Laws', *Comparative Political Studies* 41(12): 1631–56.

Garon, Francis and Éric Montpetit (2007) 'Different Paths to the Same Result: Explaining Permissive Policies in the USA', in Montpetit É., C. Rothmayr and F. Varone (eds) *The Politics of Biotechnology in North America and Europe. Policy Networks, Institutions, and Internationalization.* New York: Lexington.

Gaskell, G. and M. W. Bauer (eds) (2001) *Biotechnology 1996–2000. The Years of Controversy.* London: NMSI Trading Ltd.

Goggin, Malcolm L. and Deborah A. Orth (2004) 'The United States National talk and state action in governing ART', in Bleiklie, Ivar, Malcolm L. Goggin, and Christine Rothmayr (eds) *Comparative Biomedical Policy: Governing Assisted Reproductive Technologies.* Hove, East Sussex: Psychology Press (82–101).

Handyside A. H., E. H. Kontogianni, K. Hardy and R. M. Winston (1990) 'Pregnancies from Biopsied Human Preimplantation Embryos Sexed by Y-specific DNA Amplification', *Nature*, 344 (6268): 768–70.

Larsen, Lars Thorup, Donley T. Studlar and Christoffer Green-Pedersen (2012) 'Morality Politics in the United Kingdom: Permissiveness in a Depoliticized Context', in Isabelle Engeli, Christoffer Green-Pedersen and Lars Thorup Larsen (eds) *Morality Politics in Western Europe: Parties, Agendas and Policy Choices.* London: Palgrave Macmillan.

Linder, W. (1994) *Swiss Democracy. Possible Solutions to Conflict in Multicultural Societies.* Houndsmill: Macmillan.

Montpetit, Éric (2004) 'Policy Networks, Federalism and Managerial Ideas: How ART Non-Decision in Canada Safeguards the Autonomy of the Medical Profession', in Bleiklie, Ivar, Malcolm L. Goggin and Christine Rothmayr (eds) *Comparative Biomedical Policy: Governing Assisted Reproductive Technologies.* Hove, East Sussex: Psychology Press (64–81).

Montpetit, Éric (2007) 'A Contrast of Two Sectors in the British Knowledge Economy', in Éric Montpetit, Christine Rothmayr and Frédéric Varone (eds) *The Politics of Biotechnology in North America and Europe: Policy Networks, Institutions, and Internationalization.* Lanham: Lexington Books (103–22).

Montpetit, É. (2007) 'The Canadian Knowledge Economy in the Shadow of the United Kingdom and the United States', in É. Montpetit, C. Rothmayr and F. Varone (eds) *The Politics of Biotechnology in North America and Europe. Policy Networks, Institutions, and Internationalization.* Lanham: Lexington (83–102).

Montpetit, Éric, Christine Rothmayr and Frédéric Varone (2005) 'Institutional Vulnerability to Social Constructions: Federalism, Target Populations and Policy Designs for Assisted Reproductive Technology in Six Democracies', *Comparative Political Studies* 38(2): 119–42.

Montpetit, Éric, Christine Rothmayr and Frédéric Varone (eds) (2007) 'Comparing Biotechnology Policy in Europe and North America: A Theoretical Framework', in *The Politics of Biotechnology in North America and Europe.* New York: Lexington Books.

Papadopoulos, Y. (2001) 'How Does Direct Democracy Matter? The Impact of Referendum Votes on Politics and Policy-Making', *West European Politics*, 24(2): 35–58.

Pierson, P. (2000) 'Increasing Returns, Path Dependence, and the Study of Politics', *American Political Science Review*, 94(2): 251–67.

Pierson, Paul (2006) 'Public Policies as Institutions', in Ian Shapiro, Stephen Skowronek and Daniel Galvin (eds) *Rethinking Political Institutions: The Art of the State*. New York: New York University Press (114–31).

Ramjoué, Celina and Ulrich Köti (2004) 'Art Policy in Italy: Explaining the Lack of Comprehensive Regulation', in: Ivar Bleiklie, Malcolm Goggin and Christine Rothmayr (eds) *Comparative Biomedical Policy*. London: Routledge (42–63).

Rose, R. and P. L. Davies (1993) *Inheritance in Public Policy. Change without Choice in Britain*. Yale University Press.

Rothmayr, Christine (2003) 'Regulatory Approaches to Biomedicine: the Impact of Self-regulation on the Public Policies for Assisted Reproductive Technology', in U. Serdült and T. Widmer (eds) *Politik im Fokus, Festschrift für Ulrich Klöti*: 425–45. Zürich: NZZ Verlag.

Rothmayr, Christine (2006) 'Explaining Restrictive ART Policies in Switzerland and Germany: Similar Processes – Similar Results?', *German Policy Studies* 3(4): 595–647.

Rothmayr, Christine (2007) 'Switzerland: Direct Democracy and Non-EU Membership – Different Institutions, Similar Policies', in: Éric Montpetit, Christine Rothmayr and Frédéric Varone (eds) *The Politics of Biotechnology in North America and Europe*. Lanham: Lexington (237–61).

Rothmayr, Christine and Celina Ramjoué (2004) 'Germany: ART Policy as Embryo Protection', in Ivar Bleiklie, Malcolm Goggin and Christine Rothmayr (eds) *Comparative Biomedical Policy: Governing Assisted Reproductive Technologies*. London: Routledge (174–90).

Rothmayr, Christine and Uwe Serdült (2004) 'Switzerland: Policy Design and Direct Democracy', in: Ivar Bleiklie, Malcolm Goggin and Christine Rothmayr (eds) *Comparative Biomedical Policy: Governing Assisted Reproductive Technologies*. London: Routledge (191–208).

Rothmayr, Christine and Frédéric Varone (2009) 'Direct Legislation in North America and Europe: Promoting or Restricting Biotechnology?', *Journal of Comparative Policy Analysis: Research and Practice*, 11(4): 425–49.

Rothmayr, C., F. Varone, U. Serdült, A. Timmermans and I. Bleiklie (2004) 'Comparing Policy Design across Countries: What accounts for Variation in ART policy?', in I. Bleiklie, M. Goggin and C. R. Rothmayr (eds) *Comparative Biomedical Policy: Governing Assisted Reproductive Technologies*. London: Routledge (228–53).

Scala, F. (2003) 'Experts, embryons et "économie d'innovation": la recherche sur les cellules souches dans le discours politique au Canada', *Lien social et Politiques* 50: 75–80.

Schiffino, Nathalie, Célina Ramjoué and Frédéric Varone (2009) 'Biomedical Policies in Belgium and Italy: From Regulatory Reluctance to Policy Changes', *West European Politics*, 32(3): 559–85.

Timmermans, Arco (2004) 'The Netherlands: Conflict and Consensus on ART Policy', in Ivar Bleiklie, Malcolm L. Goggin and Christine Rothmayr (eds) *Comparative Biomedical policy. Governing Assisted Reproductive Technologies*. London: Routledge (155–73).

Timmermans, Arco (2007) 'Accommodation, Bureaucratic Politics, and Supranational Leviathan: ART and GMO Policy-Making in the Netherlands', in Éric Montpetit, Christine Rothmayr and Frédéric Varone (eds) *The Politics of Biotechnology in North America and Europe: Policy Networks, Institutions, and Internationalization*. Lanham: Lexington Books (169–92).

Timmermans, Arco and Peter Scholten (2006) 'The Political Flow of Wisdom: Science Institutions ad Policy Venues in the Netherlands', *Journal of European Public Policy* 13(7): 1104–18.

Varone, Frédéric and Nathalie Schiffino (2004) 'Regulating Biotechnologies in Belgium. Diverging Designs for ART and GMOs', *Archives of Public Health*, 62: 83–106.

Varone, Frédéric, Christine Rothmayr and Éric Montpetit (2006) 'Regulating Biomedicine in Europe and North America: A Qualitative Comparative Analysis', *European Journal of Political Research* 45: 317–43.

Part 5

Agricultural biotechnologies and the public: deliberation, opinion, ethics, and participation

11 Networkd eliberations, advocacy groups, and the legitimacy of the European Union

Éric Montpetit

Introduction

The construction of European Union (EU) institutions over the past half century has transformed democracy on the old continent. One perspective suggests that the gradual transfer of policy authority to EU institutions from member state governments empowers supranational officials, whose accountability to the European public pales in comparison with that of elected officials in member states. In contrast, another perspective suggests that EU institutions grant access to policy authority to groups of citizens that are excluded from the narrow and closed networks characterizing several policy sectors at the level of member states. Which of these perspectives is the right one? Does the EU make Europe more or less democratic? This chapter seeks answers to these questions.

The answers come from the results of a survey of biotechnology policy actors administered in the summer of 2006 and in the summer of 2008. Together, the two waves yielded 666 usable questionnaires from interest group representatives, government officials, and independent experts active in Brussels, as well as in France, the United Kingdom, the United States, and Canada. Therefore, the survey allows comparisons of actors' perceptions and attitudes between the EU and member states, but also between Europe and North America. Thanks to data on Canada and the United States, comparisons of member states to non-member states will inform assessments of the extent to which EU policy making practices have transformed democracy at the member state level.

This chapter begins with a discussion of four perspectives on democratic legitimacy. As Mansbridge writes, "When empirical political scientists want to answer the question of how well a political system meets democratic norms, they need a democratic theory that will clarify those norms in ways that make it easier to tell when real-world situations conform to or violate them" (2003: 525). This discussion will be followed by a detailed presentation of the survey. Lastly, I propose a series of arguments on the extent to which survey results match or fail to match democratic norms in the EU, member states, and non-EU countries.

Overall, survey results indicate that the EU has not made Europe any more or less democratic. First, confidence in the capacity of conventional institutions, including elections, to legitimize policy choices is not any lower at EU level than it is in nation states. Second, policy making in policy networks does not appear to be any more or less common at EU level than in nation states. Third, the EU displays more openness to new advocacy groups than North American countries. However, certain member states, France in particular, are even more open to policy deliberations involving these advocacy groups. In other words, the EU has not deteriorated democracy in Europe, but it has not improved it a great deal either.

Democratic norms

Identifying democratic norms is a prerequisite to any assessment of the quality of democracy. Mansbridge (2003) does a great favor to empirical political scientists by delineating the various norms arising from four conceptions of representation, variously prevailing in democratic states. She distinguishes between the promissory, the gyroscopic, the surrogate, and the anticipatory forms of representation, each associated with different democratic norms.

The promissory form of representation is held instinctively by most citizens. It simply holds that citizens select candidates during elections based on the promises that they make. Once elected, officials are expected to act on those promises. And when elected officials fail to do so, they are sanctioned in the following election. Here, the democratic norm is promise keeping.

The promissory conception is deficient in at least one significant way. The number of issues on which elected officials have to make decisions during a term is always significantly larger than the number of issues on which candidates make commitments during elections. Therefore, if one assumes that citizens judge politicians only through the promissory conception of representation, one also assumes that most policy decisions are irrelevant to citizens. Clearly, citizens do not focus exclusively on decisions arising from electoral campaigns.

The gyroscopic conception of representation provides an alternative way for citizens to judge politicians. This conception conceives of the election as the moment when citizens select the best candidates to govern them in the future. This conception contrasts with the promissory conception in which elections are occasions to sanction officials who failed to keep their promises in the past. In the gyroscopic conception, elections are occasions for candidates to debate in ways that reveal their respective values, ideals, and attitudes. It is not so much the promises candidates make during electoral competitions that matters to voters, as it is candidates' attitude during this competition. Voters choose candidates who have displayed an attitude that inspires confidence in their ability to govern and make acceptable decisions over a wide range of issues. In other words, elections authorize the extension of the qualities displayed by the elected candidates during the campaign to the entire term.

The promissory and gyroscopic conceptions of representation grant elections a prominent role in the creation of democratic legitimacy. To the extent elections enable sanctions in the promissory conception, and debates or deliberation in the gyroscopic conception, the quality of democracy is guaranteed. In several countries, however, electoral participation is in decline. Moreover, several elections, including EU's parliamentary elections, fail to interest voters sufficiently for a majority of them to turn out. Does this mean that such jurisdictions, including the EU, are condemned to suffer democracy deficits? Mansbridge's anticipatory and surrogate conceptions of representation insist on non-electoral sources of democratic legitimacy.

The anticipatory conception suggests that citizen's attention to democratic government is not limited to electoral periods, but also includes electoral terms. Moreover, the anticipatory approach conceives of citizens not as passive holders of preferences, but as reflexive about the various policy options facing policymakers. Therefore, representatives must constantly anticipate the evolution of citizens' preferences for the next election. Representatives have various means at their disposal to do so, including focus groups and opinion polls. However, anticipation frequently involves representatives trying to "educate" citizens about their preferred policy options. And in this educational endeavor, representatives compete with political parties, the media, and interest groups that also seek to influence citizens' preferences. Mansbridge insists on the role of interest groups in particular, which in this exercise can rectify imbalances in the relationship between representatives and citizens (see Mansbridge 2003: 520; 1992). In fact, Mansbridge argues that representatives are in 'continuing communication' over policy options with interest groups, who have better access and information than individual citizens. In any case, democratic legitimacy arises from interactions between representatives, interest groups, and ultimately citizens, to the extent that these interactions possess deliberative qualities. Deliberative qualities include the absence of coercion as well as influence exerted only through the provision of arguments on the merits of policy options. Representatives who can educate through deliberation, but who are also open to being educated by interest groups and citizens, and who can thus change their opinions, as representatives, can expect legitimacy, according to the anticipatory conception.

Surrogate representation offers an interesting complement to anticipatory representation. Valuing deliberation as much as the anticipatory conception, surrogate representation insists on the limitations created by territoriality. Most democratic institutions organize representation territorially, with particular representatives representing the citizens of given territories. However, some identities and policy positions transcend territories. Therefore, citizens occasionally feel better represented by people with whom they have no accountability relationships and feel ill represented by those who are accountable to them. For example, a woman, whose representative is male, may feel best represented by the female representative of a neighboring constituency. To make up for the deficiencies of territorial representation, interest groups can play an important role again. In fact, groups typically represent interests that transcend territories. They can provide voice for people (such as children) and even objects (such as nature), which fall outside the conventional accountability relationships organized in political institutions (Halpin 2006). In other words, interest groups can bring into deliberation perspectives overlooked by the representatives of territories, which are nonetheless significant in the pursuit of the common good. As Mansbridge writes, "the deliberative aims of democracy require that the perspectives most relevant to a decision be represented in key decisions" (2003: 524). Therefore, interest groups play a role in the production of democratic legitimacy, to the extent that they supply relevant perspectives that are frequently overseen by conventional political institutions (Bäckstrand 2006; Mansbridge 1992).

In short, the quality of democracy can be assessed through four democratic norms that are not necessarily mutually exclusive: the keeping of electoral promises; the quality of debate and deliberation during elections; the quality of deliberation between policymakers, interest groups, and citizens during terms; and the inclusion of neglected but nonetheless important groups in policy deliberation.

EU institutions and democratic norms

EU scholarship is divided between those who believe that the EU suffers from a democratic deficit and those who believe it does not (Follesdal 2006). The most pessimistic among the former group of scholars will often go as far as suggesting that EU institutions deteriorate the quality of democracy in member states (e.g. Scharpf 1999). In contrast, the most optimistic among the latter group argue that continental integration contributes positively to overall democracy in Europe (e.g. Kohler-Koch 1996).

Interestingly, those who complain about a democratic deficit will often focus their analysis on the weakness of the European Parliament and the related disinterest of citizens in European policy issues and elections (Follesdal and Hix 2005). In other words, their assessment of the quality of European democracy rests on the promissory and the gyroscopic conceptions of representation. Elections are deficient or insufficiently authorizing and therefore the quality of democracy in the EU is inferior to that of nation states. In fact, to the extent that the autonomy of democratic institutions in member states is severely constrained by EU institutions in several sectors, the democratic deficit spills over the supranational level into member states.

In contrast, optimistic EU scholars frequently have the anticipatory and the surrogate conceptions in mind (Héritier 1999; Skogstad 2003). They stress that the relative absence of hierarchical authority at the EU supranational level encourages deliberation within horizontal networks. With formal authority absent, deliberation and resulting consensus are the privileged means to make decisions. In contrast, coercion would be more common in the hierarchical settings of nation sates. They also argue that EU institutions are open to new and previously neglected perspectives (Kohler-Koch 1998). In the corporatist settings and closed policy networks of several member states, perspectives other than those of employers and trade unions would be frequently excluded from policy deliberations. In contrast, the EU would grant access to a more diversified civil society. Environmentalists, consumers, and human rights groups, to name just a few examples, would receive a fairer hearing at the EU level than within several member states.

Is it true that the EU fails so badly to meet the democratic norms of the promissory and gyroscopic models? Do national elections provide more democratic legitimacy than the election of the EU Parliament? Is it equally true that the EU does better than member states with the anticipatory and surrogate standards? Are EU decisions made more frequently in horizontal networks than decisions in nation states? Do EU decisions rest on deliberations and those of nation states on coercion? And, are those EU deliberations more inclusive of a diversity of groups and perspectives than member state deliberations, to the extent the latter exist? The rest of this chapter addresses these questions. I do so using a survey of actors who participated in the development of biotechnology policies at the EU level, in member states, and in non-EU states.

The biotechnology actor survey

Before addressing the above questions directly, I will say a few words on the methodology of the survey. The survey was conducted twice, once in the summer of 2006 and once in the summer of 2008. Together, the two waves yielded 666 usable questionnaires, with eighty-six people responding in both waves. The consistency of the results in the two waves is high (Montpetit 2009). In this chapter, I only present results of the 2008 wave, as it included more questions touching on democratic legitimacy in the EU.

Survey respondents include industry groups' representatives (80 in 2008), representatives of advocacy groups (54 in 2008), independent experts (156 scholars with university affiliations in 2008), and government employees (105 in 2008, including employees of the EU Commission). All of them were involved in the development of biotechnology policy in the agri-food or human genetics sectors in Brussels (31), France (55), the UK (67), the United States (75), and Canada (168). They were first identified as potential respondents through a search in government consultation documents, the press, and the web. The first potential respondents were then contacted via email and asked to provide names of people who were also involved in biotechnology policy development. Names were searched through this method until saturation. I thus collected contact information for 1,927 potential respondents, of which 396 yielded usable questionnaires. The response rate is twenty-one percent. This rate may appear low in comparison with conventional population surveys, but one has to keep in mind that potential respondents here are professionals solicited via their professional email address to fill out a web questionnaire. The consistency between the 2006 and 2008 waves provides some reassurance regarding the reliability of the results. Moreover, the distribution of the 396 respondents by country and by actor category is consistent with the distribution of the 1,927 potential respondents, providing additional reassurances.

This survey is unique in that it covers three categories of actors typically involved in policy networks. Most surveys of policy professionals cover only interest groups, only civil servants, or only experts, but rarely the three categories at once (one exception is Aerni and Bernauer 2005). Focusing on a single profession may be less appropriate, as network studies tell us that policy development is never the purview of a single actor category (Montpetit 2005). The survey also provides unique comparative possibilities. It enables comparison between the EU (Brussels) and two member states (France and the UK), but also with two states outside the EU: the United States and Canada. These comparisons are useful here. In fact, any failure to identify differences between the EU and the member states could be explained by the influence of the former on the latter. For example, EU decisions might proceed by deliberation, thanks to the absence of hierarchical authority. Over the years, however, this practice might have spread to member states, even though they still have hierarchical structures. Toke and Marsh (2003) provide a good example of how British biotechnology networks were transformed under the influence of the EU. A comparison with non-EU hierarchically structured states thus appears relevant.

In addition, the two sectors, agri-food biotechnology and human genetics, increase the variation in EU involvement among French and British respondents. In fact, respondents who claim that their primary involvement is in the development of French and UK policies can also be marginally involved in the development of EU policy. Clearly, however, respondents who are active in the agri-food sector are more likely to participate in EU policy development, as EU involvement in this sector is larger than in human genetics. In fact, all European respondents were asked how much of their professional time is spent working on EU policy. Results are clear: on average, agri-food respondents spend almost twenty-five percent of their time on EU policy, while less than ten percent of the time of their colleagues working on human genetics is spent on EU policy. These variations are interesting in that they enable comparisons between member state actors potentially influenced by the EU and those who are largely isolated from EU influence. Moreover, biotechnology (GMO applications in particular) regulation is a salient policy issue in Europe and therefore problems of democratic deficits in this area cannot be mistaken for the simple citizen's disinterest for marginal issues (see Morvacsik 2002). As McKay (2000) argues, citizens' judgment of a polity will more frequently stem from what the polity actually does or fails to do rather than from its institutions. In other words, a focus on specific policy issues such as biotechnology applications and on actors directly involved in this area present real advantages in assessing the EU's legitimacy.

Elections

Elections play a key role in Mansbridge's promissory and gyroscopic conceptions of representations. Promises are formulated during electoral campaigns and elections authorize some candidates to implement their promises. In the gyroscopic conception, campaigns enable debates that reveal the values and attitudes of candidates, and elections authorize some of them to govern along their values and with the attitude they displayed. In both conceptions, elections are the key devices whereby democratic legitimacy is produced. Are EU elections deficient or do they produce as much legitimacy as elections in nation states?

Arguably, citizens are better placed than interest group representatives, experts, and civil servants to answer this question. Nevertheless, group representatives, experts, and civil servants are not elected officials and therefore I have no a priori reason to believe that their judgment is significantly different from the judgment of citizens in general. In fact, if differences are there, they would possibly stem from the information actors possess about policy processes in comparison with citizens, who sometimes pay little attention to politics (Hibbing and Theiss-Morse 2002). The opinion of actors might thus be particularly revealing, as it is likely to be better informed than that of other citizens.

The survey asked respondents to rank the factors that they believed contributed most to legitimizing policy decisions in their country, or in the EU for the Brussels' respondents. The factors were:

- the scientific information upon which the decisions rest,
- the electoral mandate of the people making the decisions,
- the consultations undertaken by the decision-makers with interest groups,
- the concordance between the decisions and the preferences of the public, and
- the inherent quality of the decisions.

The results are presented in Figures 11.1 and 11.2.

Figure 11.1 shows clearly that, for most respondents, the scientific foundation of decisions is the factor contributing most to policy legitimacy. This is unsurprising given the scientific nature of the sector, biotechnology, and given that several of the respondents have scientific backgrounds. More surprising is the fact that elections are the least important legitimizing factor in the two North American countries. Moreover, elections appear more important for Brussels respondents than for

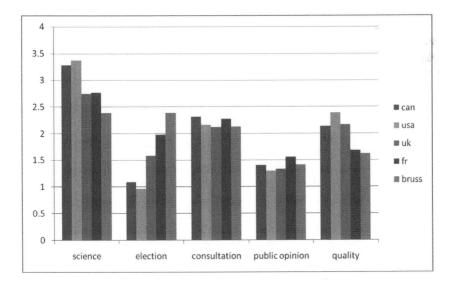

Figure 11.1 Comparison of the most important factors conferring legitimacy to policy decisions

Source: Montpetit Biotechnology Actor Survey, 2008

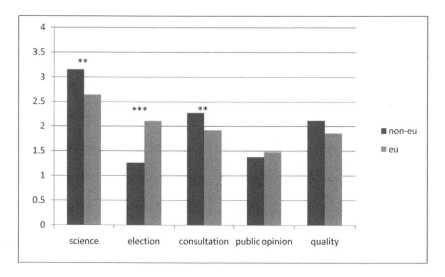

Figure 11.2 Most important legitimizing factor for actors involved in the EU and actors involved in nation states exclusively

***p. non-eu ≠ eu < 0.01; ** p. non-eu ≠ eu < 0.05 in hypothesis test of means.

Source: Montpetit Biotechnology Actor Survey, 2008

French and British respondents. In fact, for Brussels respondents, elections are just as important as science as a legitimizing factor, a finding that goes against arguments suggesting that expertise plays a relatively larger role in European policy making than in nation states (Verdun 1999). This result may be a reflection of the extensive work undertaken by the European Parliament on GMOs in recent years, which has had no parallel in the Canadian Parliament and in the American Congress. Moreover, eighty-seven percent of the respondents from Brussels belong to the agri-food sector, while Canadian and American respondents are more evenly distributed between the two sectors (seventy percent in agri-food biotechnology).

Figure 11.2 presents information that partly corrects for the bias in the distribution of agri-food respondents across jurisdictions. The figure compares the fifty-four respondents whose involvement with the EU occupies at least twenty-five percent of their professional time, with the 342 respondents whose involvement with the EU is non-existent or under twenty-five percent. In other words, EU respondents in Figure 11.2 include all respondents from Brussels, whose perception is presented in Figure 11.1, but also those French and British respondents who spend a considerable amount of their time working with EU institutions and committees. Among those respondents, seventy-four percent belong to the agri-food sector, which is closer to the North American proportion of agri-food respondents. Despite the correction in the imbalance of respondents between the two sectors, the results of Figure 11.2 are consistent with those of Figure 11.1: elections remain a more important legitimizing factor for respondents involved with the EU than for those who are not.[1]

Together, Figures 11.1 and 11.2 suggest that EU scholars, who insist on the weakness of the European Parliament and on the related deficiency of European elections, assume that the promissory and the gyroscopic conceptions of representation are more important in nation states than they are in reality. This argument is consistent with Moravscik's (2002) suggestion that concern over the EU's democratic deficit stems from a comparison with democratic ideals rarely met in nation states. As illustrated by the partisan discussion of GMOs in Canada and in the United States, some issues are never raised during nation states' electoral campaigns and, therefore, fail to become the object of promises. Some issues even fail to make it onto the agenda of parliaments, and thus they are not handled by candidates whose values and attitudes were endorsed by the electorate. Rather, a large number of issues are handled within informal policy networks. It is, therefore, unsurprising that consultation with groups emerges as a relatively important legitimizing factor on both continents in Figures 11.1 and 11.2.

Horizontal networks versus hierarchical settings

The anticipatory conception accords more importance to interactions between policymakers and civil society, especially interest groups, than to elections. Influenced by this conception, EU scholars have paid much attention to the networks of actors involved in policy development, sometimes arguing that legitimacy in the EU derives from these networks rather than from conventional political institutions (Skogstad 2003; Héritier 1999; Montpetit 2009). Because the EU's formal institutions are deficient, policy networks in the EU would have a more significant role than policy networks in nation states (Peterson 1995; 1997; Kohler-Koch 1996). To my knowledge, however, no one has ever attempted to compare the importance of networks in the EU and in nation states.

One question in the survey asked respondents whether they agreed that biotechnology policy decisions in their country (or at the EU level for Brussels respondents) were made through discussions in actor networks rather than through the exercise of authority in a hierarchical setting. In Figure 11.3, total agreement is represented by a score of five and total disagreement by a score of zero. Therefore, scores above three indicate that a majority of respondents perceived that decisions were made in networks, and scores below three indicate that a majority of respondents believed rather that decisions were made in hierarchical settings. The bars on the left of Figure 11.3 present different jurisdictions (countries and Brussels). The bars on the right compare respondents whose EU involvement is above twenty-five percent of their professional time, with those whose EU involvement is below twenty-five percent.

Figure 11.3 fails to provide a clear sense of whether decisions are made in networks or in hierarchies. In fact, every jurisdiction scores near three, and differences are not statistically significant. Moreover, respondents highly involved with the EU do not display any statistically significant difference in their inclination to perceive EU decisions as arising from policy networks. Coefficients of variation for every single score in Figure 11.3 are around thirty percent, meaning that the perceptions vary extensively from one respondent to the next. From Figure 11.3, we know that these

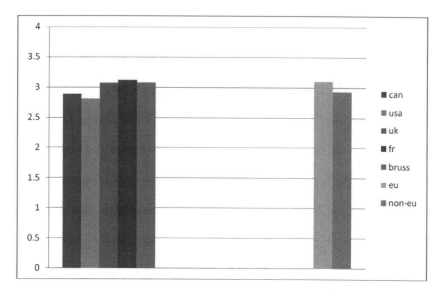

Figure 11.3 Agreement with the suggestion that decisions are made in networks rather than in hierarchical settings

***p. non-eu ≠ eu < 0.01; ** p. non-eu ≠ eu < 0.05 in hypothesis test of means.

Source: Montpetit Biotechnology Actor Survey, 2008

differences are unrelated to the origin of respondents or their involvement with EU institutions. Tests of mean differences show that they are neither related to the sector of involvement, agri-food biotechnology, or human genetics. Mean differences are only weakly related to actor category, with advocacy groups scoring clearly below three (2.76 at p = 0.07), but displaying a large coefficient of variation (forty-one percent). A possible explanation for this latter difference may be that advocacy groups are more frequently excluded from networks than the other actors, hence their inclination to perceive decisions as resting on authority (I revisit this issue below).

In sum, Figure 11.3 fails to confirm or deny that networks prevail over hierarchy. However, it leaves little doubt that EU policy actors are no more likely to perceive networks as substitutes to conventional political institutions than policy actors in member states, as well as in non-EU nation states.

Deliberation

The anticipatory conception of democratic legitimacy not only insists on the networks of policymakers, interest groups, and other actors, it also stresses the importance of the quality of the interactions among these actors. When policies result primarily from deliberative interactions between policymakers and interest groups, Mansbridge (1992) argues, we can be confident that policy choices serve the public interest. Quality deliberation requires that policymakers be inclined to adapt their preferences in the face of persuasive arguments rather than exercise coercion to meet their ends.

Some scholars suggest that deliberation (also referred to as communicative action) is particularly prevalent in the EU (Risse-Kappen 1996; Egeberg, Schäefer, and Trondal 2003; Joerges and Neyer 1997). The weakness of conventional institutions prevents authoritative arbitration among policy positions. The Council of Ministers, the Commission, and the Parliament prefer making decisions on issues over which consensus was formed in committee meetings (Wessels 1998; Costa 2000; Lijphart 1999). And consensus is all the more likely to succeed where actors have preferences that can change in the face of convincing arguments. Otherwise, attempts to reach consensus risk ending in deadlock. Again, deliberation and consensus could make up for the weakness of conventional political institutions in the EU. Summarizing the literature on EU governance, Pollack writes, "students of EU governance often (although not always) emphasize the capacity of the EU to foster 'deliberation' and 'persuasion' – a model of policy-making in which actors are open to changing their beliefs and their preferences, and in which good arguments can matter as much as, or more than, bargaining power" (2005: 36).

Does the EU encourage deliberation to a larger extent than nation states? One question in the survey asked whether people making biotechnology policy decisions in respondents' respective countries (or the EU for Brussels respondents) can be persuaded to change their mind when faced with convincing arguments. Results are presented in Figure 11.4. Here again, three is the tipping point between agreement and disagreement.

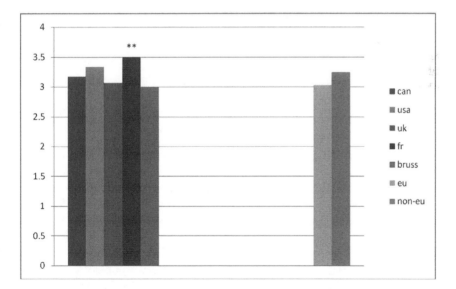

Figure 11.4 Agreement with the suggestion that policymakers can be persuaded to change their mind

** p. fr ≠ other respondents < 0.05 in hypothesis test of means.

Source: Montpetit Biotechnology Actor Survey, 2008

In contrast with Figure 11.3, the overall result leans toward agreement (3.22), but with a large coefficient of variation of thirty-one percent. I repeated the same test, excluding government employees who might believe that they count as policymakers and therefore might overestimate their capacity to alter their opinion. The results remained similar.

Brussels respondents (on the left side of Figure 11.4) and respondents most involved with the EU (on the right side) are, on average, more inclined to disagree that policymakers can change their mind. The statistical significance of the difference, however, is very weak (p. non-eu ≠ eu = 0.16). Even more interesting is that, of all respondents, the French are most inclined to agree that policymakers change their minds in the face of persuasive arguments and the difference is statistically significant.[2] In a previous article, a colleague and I argued that the French risk management culture for biotechnology creates a predisposition favorable to debates and presumably deliberation (Montpetit and Rouillard 2008). Figure 11.4 adds evidence of the existence of this culture. It specifies the strength of its distinctiveness in comparison with North American countries, but also in comparison with the United Kingdom and Brussels. Figure 11.4 clearly shows that the French deliberative culture is unique and largely unrelated to Europeanization.

In short, the anticipatory conception insists on network deliberation as a source of legitimacy. If indeed network deliberation matters, it matters equally in most nation states, member states, and Brussels, despite the EU's weak political institutions. In fact, governance by network does not appear any more common in the EU level than elsewhere. Moreover, deliberation distinguishes France, a country where political institutions centralize policy authority perhaps indicating that deliberation and consensus are easier to achieve in the shadow of hierarchy (Héritier and Lehmkuhl 2008).

Inclusiveness of perspectives

The surrogate conception insists on the inclusion of a diversity of perspectives during policy deliberations. Based on territoriality, conventional institutions can represent citizens inadequately, hence the importance of granting a policy making role to groups voicing the preferences of marginalized citizens. In addition, some analysts have complained that the relatively closed networks prevailing in some European countries have excluded too many perspectives from policy making. Thanks to the EU, empirical analyses suggest, the representatives of these perspectives have had opportunities to participate in policy deliberation in recent years (Toke and Marsh 2003; Bernauer and Meins 2003). Does the EU, in fact, enable the inclusion of neglected perspectives in policy making?

Biotechnology is a particularly appropriate sector to answer this question. It is a relatively recent sector, which is not characterized by the traditional opposition between labor and capital. Industry surely plays a large role in this area, but it is confronted by advocacy groups with environmental, health and ethical concerns. The perspectives of these latter groups are good examples of perspectives presumably excluded from the corporatist network of member states. Therefore, any evidence that their representatives receive more hearing time at the EU level than at member state level would justify a defense of the EU's legitimacy in terms of the surrogate conception of representation.

One of the survey questions asked whether respondents agreed that biotechnology policy decisions in their respective countries (or EU for Brussels' respondents) are preceded by deliberation with a wide range of actors. A higher degree of agreement on the part of advocacy groups involved at the EU level can be interpreted as evidence that the EU is more open to new perspectives than member states. Generally speaking, the representatives of advocacy groups are more likely than the other actors to be excluded from policy networks, hence the relevance of their opinion about the inclusiveness of perspectives in policy making. In fact, agreement by advocacy groups that policy decisions are preceded by deliberation with a wide range of actors should be a good indicator of inclusiveness. The results are presented in Figure 11.5 and the tipping point between agreement and disagreement is three, as in the previous figures.

The results of Figure 11.5 have to be interpreted with care as the numbers are insufficiently high for mean differences to be statistically significant (they were calculated from the fifty-four representatives of advocacy groups only). In light of the previous figures, it is notably strange that France scores so low. In any case, Figure 11.5 indicates that representatives of advocacy groups in Brussels more

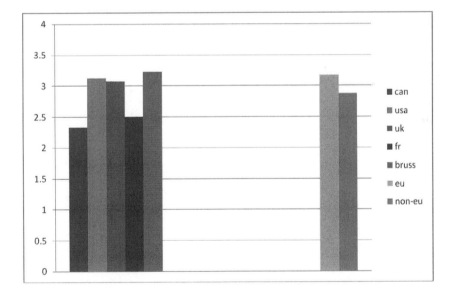

Figure 11.5 The extent to which advocacy groups agree that policy decisions are preceded by deliberation with a wide range of actors

***p. a jurisdiction ≠ others < 0.01; ** p. a jurisdiction ≠ others < 0.05 in hypothesis test of means.

Source: Montpetit Biotechnology Actor Survey, 2008

frequently believe that policy deliberations involve a wide range of actors than advocacy groups elsewhere.

Additional survey questions can make up for the statistical deficiency of Figure 11.5. For example, one question asked respondents to indicate how frequently they meet individuals or groups whose views on biotechnology policy differ from their own. The frequency whereby industry representatives, experts, and government employees meet with actors holding different views could provide an indication of openness to deliberation with advocacy groups (Mutz 2006). In fact, industry representatives, experts, and government employees are more likely to be mainstream actors. The results are presented in Figure 11.6. Zero corresponds to never; one corresponds to about once a year; two to about once every six months; three to about once a month; and four to more than once a month. Results regarding the EU are consistent with Figure 11.5: respondents from Brussels meet people with different views more frequently.[3] Interestingly, the results for France in Figure 11.6 also confirm previous results suggesting that deliberative attitudes are strong in this country.

Naturally, the survey measures the attitude of respondents toward biotechnology. Therefore, it enables measures of the extent to which the apparent openness of the EU to advocacy groups translates into a wider range of perspectives being represented during policy deliberations. In fact, a series of questions asked whether respondents agreed on the risks and benefits of a range of biotechnology applications. Whether respondents agreed that GMOs are harmful to organic production is an example of a question insisting on risks. Whether respondents agreed that GMOs can improve public health is an example of a question insisting on benefits. Using four questions

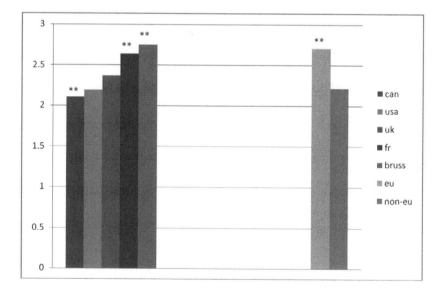

Figure 11.6 Frequency whereby industry, experts and government employees meet with people holding different views

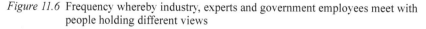

** p. a jurisdiction ≠ others < 0.05 in hypothesis test of means.

Source: Montpetit Biotechnology Actor Survey, 2008

insisting on risks, I constructed a measure (between 0 and 10) of precautionary attitudes towards biotechnology. Similarly, I used four questions insisting on benefits to construct a measure (between 0 and 10) of biotechnology promotional attitudes. At 7.14, the commitment to promotion is significantly higher than the commitment to precaution, which averages 5.57. Advocacy groups, however, score much higher than the other actors on precaution, averaging 7.56 against 5.24 for promotion. In other words, advocacy groups are more marginal in their precautionary views than the other mainstream actors, who prefer promotion.

It follows that ratio estimates of precautionary versus promotional attitudes by jurisdiction provide measures of the extent to which advocacy groups are heard in their respective jurisdictions. The closer the ratio is to one, the more influential the precautionary view, which is most vigorously defended by advocacy groups. The further it is under one, the less influential the precautionary view, indicating that advocates of promotion prevail over most advocacy groups. The ratio estimates are presented in Figure 11.7.

Figure 11.7 shows that the United States fail to represent advocacy groups to the same extent as the other jurisdictions. Respondents may feel that policymakers consult with a wide range of groups, as indicated in Figure 11.5, but promotional attitudes prevail over precautionary attitudes, in comparison with other jurisdictions. In contrast with Figure 11.6, results for Canada are closer to European rather than American results. Otherwise, Figure 11.7 is consistent with Figure 11.6: advocacy groups are better represented and more influential in Europe than in North America.

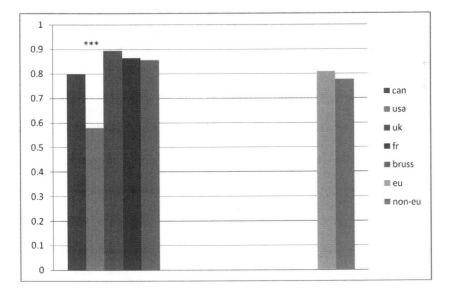

Figure 11.7 Ratio estimations of precautionary/promotional attitudes toward biotechnology

***p. a jurisdiction ≠ others < 0.01 in hypothesis test of means.

Source: Montpetit Biotechnology Actor Survey, 2008

However, the better representation of advocacy groups' views in Europe appears unrelated to the EU. According to Figure 11.6, meetings between mainstream actors and advocacy groups are as common in Paris as they are in Brussels. According to Figure 11.7, precautionary perspectives are just as well represented in London and Paris as they are in Brussels. Moreover, the right side bars of Figure 11.7 shows that precautionary views are not any better represented among those most involved with the EU. In short, the EU can claim legitimacy along the surrogate conception of representation. However, the EU cannot be legitimized with suggestions that it makes up for the narrowness of the perspectives included in member states' policy networks.

Conclusion

Does the EU make Europe more democratic? The evidence presented in this chapter clearly suggests a negative answer. However, this negative answer is different from any suggestion that the EU suffers a legitimacy deficit. As Moravscik (2002) argues, the ideal of democratic legitimacy is rarely met in nation states in the first place (see also Héritier 1999). In fact, in the biotechnology sector, the Canadian and American legislative assemblies were much less involved than the European Parliament. Biotechnology might not have been the object of electoral promises during European elections, but electoral debates and elections helped distinguish between candidates who inspired trust and those who did not. Therefore, the work of elected European officials on biotechnology received some legitimacy, according to a gyroscopic conception of representation. This is more than can be said about the relative absence of legislative debates on GMOs in Canada and in the United States.

Conversely, EU scholars who insist on network deliberation as an alternative to legitimization through conventional political institutions in the EU, might underestimate its importance in member states and non-EU nation states (e.g. Skogstad 2003). In fact, as Mansbridge (2003) argues, the anticipatory conception of representation frequently makes up for the deficiencies of the promissory and gyroscopic conceptions, which afflict most nation states. In other words, network deliberation might help with legitimizing the EU, but it is equally if not more important in member states and non-EU countries.

In comparison with North America, however, Europe appears more inclusive of marginal voices in biotechnology policy debates. This observation translates certainly into the production of more legitimacy through the surrogate conception of representation in Europe than in North America. An additional consequence is the development of European biotechnology policies that are frequently more restrictive than comparable policies in North America (Bernauer and Meins 2003; Montpetit et al. 2007). However, the inclusion of advocacy groups in policy deliberations is unlikely a result of Europeanization. Brussels does not appear to be a source of inspiration in terms of inclusiveness, as London and Paris are just as inclusive of advocacy groups and of its precautionary perspective. A longitudinal study, however, would provide stronger evidence regarding this latter conclusion.

When the results for all four conceptions of representation are added, the legitimacy of Europe's biotechnology policy appears slightly ahead of that of North America. Again, this result appears unrelated to Europeanization. In other words, EU institutions did not add or remove anything to member states' capacities to make legitimate policy decisions.

Notes

1 The difference remains in regression analyses (poisson and negative binomial) using gender, sector, actor categories and years of involvement as control variables. I also conducted the tests without government employees and the results were similar.
2 However, it loses its significance in a poisson regression using the control variables listed in note 1.
3 Differences hold in a poisson regression.

Bibliography

Aerni, P. and Bernauer, T. (2005) 'Stakeholder Attitudes toward GMOs in the Philippines, Mexico, and South Africa: The Issue of Public Trust', *World Development*, 34(3): 557–75.

Bäckstrand, K. (2006) 'Multi-Stakeholder Partnerships for Sustainable Development: Rethinking Legitimacy, Accountability and Effectiveness', *European Environment*, 16: 290–306.

Bernauer, T. and Meins, E. (2003) 'Technological Revolution Meets Policy and the Market: Explaining Cross-National Differences in Agricultural Biotechnology Regulation', *European Journal of Political Research*, 42: 643–83.

Costa, O. (2000) *Le Parlement européen, assemble délibérante*. Brussels: Éditions de l'Université de Bruxelles.

Egeberg, M., Schäefer, G.F. and Trondal, J. (2003) 'The Many Faces of EU Committee Governance', *West European Politics*, 26: 9–40.

Follesdal, A. (2006) 'Survey Article: The Legitimacy Deficit in the European Union', *The Journal of Political Philosophy*, 14: 441–8.

Follesdal, A. and Hix, S. (2005) 'Why There Is a Democratic Deficit in the EU: A Response to Majone and Moravcsik', *European Governance Paper* No. C-05-02.

Halpin, D.R. (2006) 'The Participatory and Democratic Potential and Practice of Interest Groups: Between Solidarity and Representation', *Public Administration*, 84: 919–40.

Héritier, A. (1999) 'Elements of Democratic Legitimation in Europe: An Alternative Perspective', *Journal of European Public Policy*, 6(2): 269–82.

Héritier, A. and Lehmkuhl, D. (2008) 'Introduction: The Shadow of Hierarchy and New Modes of Governance', *Journal of Public Policy*, 28: 1–17.

Hibbing, J.R. and Theiss-Morse, E. (2002) *Stealth Democracy: Americans' Beliefs about How Government Should Work*. Cambridge: Cambridge University Press.

Joerges, C. and Neyer, J. (1997) 'Transforming Strategic Interaction into Deliberative Problem-Solving: European Comitology in the Foodstuffs Sector', *Journal of European Public Policy*, 4: 609–25.

Kohler-Koch, B. (1996) 'Catching Up with Change: The Transformation of Governance in the European Union', *Journal of European Public Policy*, 3(3): 359–80.

Kohler-Koch, B. (1998) 'Organised Interests in the EU and the European Parliament', in P. H. Claeys, C. Gobin, I. Smets and P. Winand (eds) *Lobbying, Pluralism and European Integration*. Brussels: European University Press.

Lijphart, A. (1999) *Patterns of Democracy: Government Forms and Performance in Thirty-Six Countries*. New Haven: Yale University Press.

Mansbridge, J. (1992) 'A Deliberative Theory of Interest Representation', in M. Petracca (ed.) *The Politics of Interest: Interest Groups Transformed*. Boulder: Westview Press.

Mansbridge, J. (2003) 'Rethinking Representation', *American Political Science Review*, 97: 515–28.

McKay, D. (2000) 'Policy Legitimacy and Institutional Design: Comparative Lessons for the European Union', *Journal of Common Market Studies*, 38: 25–44.

Montpetit, É. (2005) 'A Policy Network Explanation of Biotechnology Policy Differences between the United States and Canada', *Journal of Public Policy*, 25(3): 339–66.

Montpetit, É. (2009) 'Governance and Policy Learning in the European Union: A Comparison with North America', *Journal of European Public Policy*, 16(8): 1185–203.

Montpetit, É. and Rouillard, C. (2008) 'Culture and the Democratization of Risk Management: The Widening Biotechnology Gap between Canada and France', *Administration & Society*, 39: 907–30.

Montpetit, É, Rothmayr, C. and Varone, F. (eds) (2007) *The Politics of Biotechnology in North America and Europe: Policy Networks, Institutions and Internationalization*. Lanham: Lexington Books.

Moravscik, A. (2002) 'In Defense of the Democratic Deficit: Reassessing Legitimacy in the European Union', *Journal of Common Market Studies*, 40: 603–24.

Mutz, D.C. (2006) *Hearing the Other Side: Deliberative versus Participatory Democracy*. Cambridge: Cambridge University Press.

Peterson, J. (1995) 'Policy Networks and European Union Policy Making: A Reply to Kassim', *West European Politics*, 18: 289–407.

Peterson, J. (1997) 'States, Societies and the European Union', *West European Politics*, 20(4): 1–23.

Pollack, M.A. (2003) 'Control Mechanism or Deliberative Democracy? Two Images of Comitology', *Comparative Political Studies*, 36: 125–55.

Pollack, M.A. (2005) 'Theorizing EU Policy Making', in H. Wallace, W. Wallace and M.A. Pollack, (eds) *Policy Making in the European Union, fifth edition*. Oxford: Oxford University Press.

Risse-Kappen, T. (1996) 'Exploring the Nature of the Beast: International Relations Theory and Comparative Policy Analysis Meet the European Union', *Journal of Common Market Studies*, 34(1): 53–80.

Scharpf, F. (1999) *Governing in Europe: Effective and Democratic?* Oxford: Oxford University Press.

Skogstad, G. (2003) 'Legitimacy and/or Policy Effectiveness? Network Governance and GMO Regulation in the European Union', *Journal of European Public Policy*, 10: 321–38.

Toke, D. and Marsh, D. (2003) 'Policy Networks and the GM Crops Issue: Assessing the Utility of a Dialectical Model of Policy Networks', *Public Administration*, 81: 229–51.

Verdun, A. (1999) 'The Role of the Delors Committee in the Creation of EMU: An Epistemic Community?', *Journal of European Public Policy*, 6(2): 308–28.

Wessels, W. (1998) 'Comitology: Fusion in Action. Politico-Administrative Trends in the EU System', *Journal of European Public Policy*, 5(2): 209–34.

12 Getting to maybe

Assessments of benefits and risks in Canadian public opinion on biotechnological innovation

Steven Weldon, David Laycock,
Andrea Nüsser and Colin Whelan

Introduction

The burgeoning fields of biotechnology and genomic science have seen rapid advances in recent years with a host of new applications in medicine, agriculture, and related fields. Advocates and scientific experts point to the revolutionary potential of these technologies to improve human health, the environment, and the overall quality of life. This potential for radical change has, however, also raised public concerns. Indeed, opinion surveys in advanced industrial democracies consistently reveal sceptical publics that are uncertain about the technologies' long-term safety, uneasy about the ethical implications, and doubtful of the purported benefits – particularly as all of these relate to genetic modification. Public support from both consumers and democratic citizens is critical to the future viability and development of new biotechnologies.

The importance of public opinion to governments' and public responses to new biotechnologies, and hence to their future, has led to a growing and diverse social science research agenda over the last three decades concerning the determinants of public support for biotechnologies. The dominant approach remains the 'Deficit Model', which holds that opposition to emerging technologies stems largely from ignorance (information deficits) about their substantial benefits and insubstantial risks. Other scholars challenge the narrow explanatory focus on a simple, cost–benefit calculus and examine a larger range of 'risk' factors that may influence individual judgments. Currently, academic efforts to understand public opinion on such matters can be organised into three broad approaches. These focus on public knowledge about and attentiveness toward science, public trust in institutional actors and regulatory bodies, and the impact of citizens' values and ethical considerations on their views of biotechnology.

The broadening of the range of risk factors considered as potential influences on public judgments to include trust in institutional actors and ethical/moral concerns has helped us better understand the correlates of support for new technologies. Yet, we still do not have a good understanding of exactly how individuals make judgments – that is, which factors matter for which people and when. The assumption that citizens employ a single logic in making such judgments underlies most contemporary research in this field, regardless of researchers' competing emphases on different factors. In particular, each approach tends to assume that citizens engage in relatively *balanced* risk–benefit assessments of specific innovations, whether consciously or subconsciously, perhaps with assistance from various heuristics (e.g. Slovic *et al.* 2007).

An exception is the recent work of George Gaskell and his colleagues, which challenges this assumption, suggests that the disproportionate emphasis on risks over benefits in the literature is misplaced, and contends that multiple decision-making logics likely undergird public opinion on modern biotechnologies (Gaskell *et al.* 2004). Specifically, based on a 1999 Eurobarometer survey, they find most Europeans who reject GMO foods are sufficiently dismissive of the potential benefits that they do not bother to assess risks. That is, many respondents do not engage in a balanced risk–benefit assessment; instead, once they determine there are no or limited benefits to a technology, many respondents appear to stop their individual level deliberations before risks are assessed, and move directly to a negative verdict. Making sense of how citizens perceive both benefits and risks, or how they can succeed in 'getting to maybe', then becomes a critical step in understanding attitudes toward new technologies.

We build on this intriguing study, and other recent work, to develop a more detailed causal model that takes into account the possibility of different logics of opinion formation. The model causally separates out considerations of risks and benefits; it also suggests that considerations of 'moral risk' operate largely outside the traditional risk–benefit calculus, while trust in institutional and regulatory actors is mainly relevant for those who 'get to maybe' and perceive both risks and benefits. We then use data from a 2009 Canadian public opinion survey to test aspects of this model.

The next section provides an overview of historical support for different biotechnologies as well as results from the 2009 survey. We then briefly review previous research on predictors of support, before moving on to present our causal model, which divides respondents into four groups based on their perceptions of benefits and risks. Using the 2009 Canadian survey, we then examine predictors of perceived benefits and risks respectively in separate multiple regression models. We conclude with a discussion of our model and the implications for future research in this field.

Charting support for biotechnologies

From genetic testing for hereditary diseases, to DNA fingerprinting in criminal cases, to the introduction of genetically modified crops, modern biotechnology has transformed our understanding of science and its potential for altering human society. Genome sequencing and breakthroughs in stem cell and cloning research promise to push the frontiers of this science even further. In the medical sciences, proponents predict genomic advances will lead to new drugs and cures for many life-threatening diseases, such as cancer, heart disease, diabetes, and Alzheimer's. In the agricultural sciences, advocates contend that breakthroughs will produce crops with increased nutritional value and greater disease resistance, thus increasing production yields, helping to combat world hunger, reducing environmental damage due to pesticides, and reducing CO_2 warming through increased bio-fuel use.

Despite the obvious promise of emerging biotechnologies, obvious and widespread support by governments and the scientific community, opinion surveys have regularly found that many members of the public are sceptical of their purported benefits and uneasy about their ethical implications and long-term safety. Public support has been weakest in Europe, as evidenced by Eurobarometer surveys on science and technology (Gaskell *et al.* 2001; Durant and Legge 2005; Weldon and Laycock 2009). While support has been greater in North America, Canadians and Americans still remain divided – particularly concerning genetic modification of organisms (Gaskell and Jackson 2005).

Table 12.1 presents findings from our 2009 survey on Canadian support for different biotech applications in medicine and agriculture. The values represent the percentage of survey respondents' support for each application.[1]

Table 12.1 Canadians' approval of different biotechnology applications

Technology	Approve without reservation	Approve with usual regulation	Approve with stricter regulation	Approve only in pecials circumstances	Do not approve at all	Don't know
Test for Cancer Risk	18.2%	29.9%	26.9%	15.5%	2.3%	7.2%
GM Diabetes	8.4%	31.5%	34.0%	10.5%	5.1%	10.5%
Nanotech (Medical)	17.4%	37.2%	32.8%	5.0%	2.5%	5.2%
GM Bird Flu Vaccine	12.8%	31.0%	34.9%	11.9%	3.3%	6.0%
GM Potato	8.9%	19.6%	13.2%	22.4%	19.0%	16.9%
GM Pine Trees	11.4%	25.0%	28.8%	16.8%	3.2%	14.9%
Salmon Biomarkers	9.5%	21.1%	34.0%	12.9%	9.5%	12.9%
Wine Bio-markers (Diagnostic)	10.0%	37.9%	26.4%	8.8%	2.7%	14.2%
Wine Bio-markers (Breeding)	8.1%	37.1%	30.1%	9.6%	5.5%	9.6%
GM Wine Vines	8.7%	34.9%	24.8%	13.8%	6.1%	11.8%

Source: Angus Reid Survey, June 2009.

Consistent with previous studies, medical applications receive decidedly greater support than agricultural ones. Among the medical applications, however, several findings cause us to question whether this is significant and reliable support. First, none of the applications receive majority support based on their current level of regulation. Approval does jump markedly with tighter regulation of these technologies, but this finding indicates just how uncertain many Canadians are about their risks, even if they recognise their potential benefits. Second, the least radical but best known application, a genetic test for cancer risk, actually engenders the greatest opposition and polarisation. It is intriguing that the two medical technologies that involve genetic modification, especially the GM Bird Flu vaccine with mandatory vaccinations for the elderly and children, receive similar levels of support as genetic testing. Perhaps support for new technologies may drop when they move from hypothetical to current application (and the risks become more immediate), but it is also possible that support is greater for technologies that directly solve problems as opposed to simply identifying them.

Turning to the non-medical uses, it is the level of public opposition to GM potatoes that really stands out. Why GM wine vines are not also the object of strong opposition is not immediately clear, given the well documented opposition to GM foods, but people may see a distinction between the vine and the fruit it bears. Some Canadians may also believe that they can discover whether such products are in their wine more easily than they can with bulk potatoes purchased in grocery stores, or that basic foodstuffs are of greater concern than luxury consumption items.[2] Support is also higher for applications using biomarkers for diagnostic purposes or selective breeding,[3] and higher in the case of GM pine trees resistant to the mountain pine beetle, where a problem is well established in Canadian public discourse.[4]

Taken together, the results suggest the complexity of opinions toward new biotechnologies. The public does not appear to simply support medical uses and oppose agricultural ones, nor does it uniformly oppose genetic modification. Instead, individuals appear quite discerning in their preference formation, at least more than many scientific experts credit them to be, and consider the benefits and risks of individual technologies in their decision-making processes. In the following sections, we seek to better understand this process and develop a unified model that takes its complexity into account.

Past research on sources of support for biotechnology

Past research on the sources of support for biotechnology has utilised three main approaches. The oldest, still dominant, approach focuses on individual knowledge and attentiveness toward science and biotechnology. Commonly termed the 'Deficit Model', it holds that opposition to emerging technologies stems largely from ignorance about their benefits and an irrational fear of their risks (Evans and Durant 1995; Irwin and Wynne 1996; Sturgis and Allum 2004). Buoyed by studies repeatedly confirming the public's ignorance about science and the benefits of scientific innovation, the model also assumes individuals would embrace these technologies if they were simply more knowledgeable about their benefits and limited risks (much like the scientific community and regulators).

However, several recent studies have undermined the basic tenets of the Deficit Model. These studies, often of deliberative settings where participants actively discuss and learn about a specific technological application, suggest that greater, and more objectively accurate knowledge of the risks and benefits (but not the science behind their assessment), does not necessarily lead to greater support for the innovative application – it may even increase opposition (Hamlett *et al.* 2008; see also Gavelin *et al.* 2007). Others have found that the effect of knowledge depends on the specific technology in question (Evans and Durant 1992; Sturgis *et al.* 2005). There is also a substantial community of scholars that have long been critical of the Deficit Model's narrow focus on formal scientific knowledge, which has argued that other forms of knowledge or cues can substitute for formal scientific knowledge in helping individuals to make effective policy judgments (Irwin and Wynne 1996; Durant *et al.* 2000). Nonetheless, a key question emerges from the recognition that much of the public is 'scientifically illiterate': how *do* individuals then make judgments about science and new technologies?

The second major approach to understanding public opinion toward new technologies responds to this question by focusing on the institutional context of scientific research, including trust of regulatory actors and stakeholders (Yearley 2000; Priest *et al.* 2003; Legge and Durant 2010). Institutional actors are a key source of information about new technologies, particularly when individuals perceive a personal lack of relevant knowledge. If 'official' sources of scientific knowledge – scientists and state regulators – are not trusted in post-industrial societies (Crawley 2007), their claims are likely to be consciously or unconsciously rejected, in favour of information from relatives, friends, social or political organisations, or other perceived experts (Bennett and Calman 1999). More importantly, since most citizens implicitly charge governments and regulators with balancing the risks and benefits of new technologies to promote the public good, a lack of trust in public officials in this regard further erodes public support for new technologies.

Previous research has confirmed the importance of public trust in institutional actors for public support of new technologies (Grove-White *et al.* 1997; Siegrist 2000; Gottweiss 2002; Priest *et al.* 2003; Durant and Legge 2005). Barnett *et al.* (2007) look beyond trust in stakeholders and find that levels of trust in government rules and regulatory bodies in Great Britain are also much stronger predictors of support for genomic applications than public attentiveness to genetics and education. Priest *et al.* (2003) identify a marked 'trust gap' between competing regulatory and social group actors in Europe and the United States, and argue compellingly that the 'trust gap' is the most decisive factor in explaining variation in individual attitudes toward biotechnologies.

A second response to the Deficit Model, and the third major approach to studying public opinion toward new technologies, focuses on ethical concerns and core values. These factors include religious and moral inclinations, as well as quasi-spiritual orientations to living in a natural order/protecting nature, and post-material values regarding the overall quality of life. Recent studies of New Zealanders' attitudes towards various biotechnology applications contend that affective orientations towards specific biotechnologies, and respondents' general attitudes towards nature and technology, are central to explanations of attitudes towards biotechnologies (Fairweather and Cook 2005; Coyle and Fairweather 2005). Such orientations towards nature and technology are embedded in spiritual beliefs, worldviews, and post-material value positions. Sjoberg (2000; 2005) argues that moral considerations ground technological risk perceptions, and that public concerns about interfering with nature, the moral value of technology, and trust in science offer stronger explanatory force than affective and risk assessment factors.

A unified causal model of opinion formation

Broadening risk factors to include trust in institutional actors and ethical and moral concerns has given us a better understanding of the correlates of support for new technologies. Yet we still lack a good understanding of exactly how individuals make judgments – that is, which factors matter for which people, and when. Similarly, the assumption that there is a single, universal logic in making such judgments underlies most contemporary research, regardless of the emphasis on different factors. Lack of trust in institutional actors and moral and ethical concerns are simply treated as other forms of risk. Along with concerns about safety risks (from the Deficit Model), these factors go directly into the risk part of the presumably additive risk–benefit calculus that determines support for different new technologies. Building on Gaskell and colleagues' 2004 article, 'GM Foods and the Misperception of Risk Perception', we develop and test a model below that challenges both of these assumptions.

Gaskell and his colleagues found that most European citizens opposed to GM foods in the late 1990s did not undertake a balanced assessment of perceived benefits and risks; they either discounted or ignored risk information altogether. As they note, a response that dismisses the possibility of an innovation's benefit and skips over any considerations of potential risks when moving to a final judgment, effectively rejects the potential value of such products. From the perspective of the innovation's advocates, this is tantamount to denying the only compelling rationale for such an innovation.[5] More broadly, it indicates a need to reconsider our approach to studying opinions about new technologies. Not everyone is going to see benefits, and for them, other factors known to relate to individual support, including trust in institutional actors, will be largely irrelevant. In a similar vein, moral or ethical concerns about biotechnology are likely to override other considerations, including perceptions of benefits to human health or the environment.

Figure 12.1 lays out our initial efforts to develop a unified causal model of public opinion formation with respect to biotechnologies. The model takes into account the three major existing strands of research. It allows for different pathways of

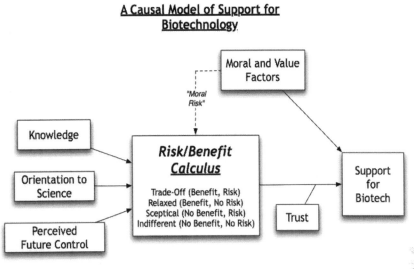

Figure 12.1 Modelling support for biotech

decision making and identifies when key factors known to correlate with support for new technologies should actually matter most in the decision-making process. We discuss this model in light of the existing literature.

At the centre of the model is the traditional risk–benefit calculus where, much like the Deficit Model, risk is only meant to capture perceived *safety* concerns about adopting a technology. However, in contrast to the Deficit Model, we see risks and benefits along two independent, separate dimensions (Poortinga and Pidgeon 2005). Most importantly, this allows for the very realistic possibility that individuals simultaneously perceive both benefits and risks in a new technology. Following Gaskell and his colleagues, and using their terminology, we classify individuals into a four-fold typology based on whether one perceives a technology's respective risks and benefits to be high or low. The groups are:

- the 'Tradeoffs' who perceive both risks and benefits;
- the 'Relaxed' who perceive benefits but no risks;
- the 'Sceptical' who see risks and no benefits; and
- the 'Indifferent' who perceive neither benefits nor risks.

A critical assumption here is that each group makes judgments about new technologies in different ways, particularly as it relates to the role of trust in institutional actors and regulators. Thus, an important first step is to understand who perceives benefits and risks. On the left side of the model, we identify three key sets of causal factors that we hypothesise affect assessments of benefits and risks: scientific knowledge (and attentiveness), general orientation to science, and perceived personal control over the future use of a technology.[6] The former two are aspects of the Deficit Model;

we also expect that increased knowledge and a positive general orientation toward science tend to increase perceptions of biotechnologies' benefits. However, following recent research discussed above, we further expect that increased knowledge leads to the perception of greater risks of these same technologies (Poortinga and Pidgeon 2005; Weldon and Laycock 2009). In other words, we hypothesise that scientific knowledge increases perceptions of both benefits *and* risks.

The third key factor we identify is perceived ability to control the personal consumption or use of a technology in the future. Opinion surveys have consistently shown greater support for medical biotech applications over agricultural ones. Gaskell and his colleagues (2004) postulate this occurs when individuals fail to see any tangible benefit in the latter and thus reject them out of hand without considering potential risks. The authors suggest that this behaviour is consistent with decision-making theory based on a 'lexicographic heuristic', which contends that top-ranked attributes or considerations generally sideline lower-ranked considerations and concerns in decision making (Fishburn 1974).

We offer an alternative explanation that focuses on the risk side. Most medical applications also offer a high degree of personal control and hence mitigate any potential safety risk; one can support development now, while still maintaining the choice to use the application in the future. In other words, these innovations are not forced on people – their use is only relevant if one has a specific ailment, and even then, one could choose to opt for more conventional treatments. Agricultural biotech applications, by contrast, offer decidedly less personal control, particularly those commonly surveyed, such as GMO foods. In the absence of labelling, there is virtually no personal control, but even with it, individuals are still quite likely to consume GMO products (inadvertently). This increases the (perceived) risk of such technologies.

Moving to the right side of Figure 12.1, we model the role of moral and normative factors, which we see as operating largely outside of the traditional cost–benefit calculation. Grounded in ideological convictions, worldviews and spiritual orientations (Fairweather and Cook 2005; Coyle and Fairweather 2005; Sjoberg 2000, 2005), these core values act as important cues for social and political decision making. In politics, such core values have been found to shape strongly more specific issue positions, even trumping individual self-interest (Zaller 1992). In the case of support for biotechnologies, one could be knowledgeable about a technology's benefits and risks, even believe that the benefits are high and the safety risks are low, yet still fail to support that technology for ethical or value reasons (Traill *et al.* 2004).

Theoretically, we expect that this is how ethical factors largely operate; when moral and ethical concerns are present, they are primarily causal in determining support for biotechnologies, and effectively trump other considerations. Empirically, however, these effects are difficult to isolate, because they may also cloud perceptions of benefits and especially risks. We thus include these 'moral' factors in our causal model in a way that accommodates both possibilities. The dashed arrow suggests they may also affect support for biotechnologies indirectly, through their effect on perceptions of benefits and risks (primarily risks).

Finally, we turn to the role of trust in institutional actors and regulators. That citizen trust in various scientific, government, or corporate elites structures support for emerging technologies is well established in the literature (Yearley 2000; Priest *et al.* 2003; Frewer *et al.* 2003; Durant and Legge 2005; Barnett *et al.* 2007; Legge and Durant 2010).

However, we diverge from previous research in two respects. First, our model indicates that citizens' trust in institutions or regulators does not directly affect perceptions of a technology's risks or benefits. Again, it is important to note that we confine risk here to just perceptions of safety risks. In this case, we expect that trust in actors only enters into the decision-making process *after* determining risks and benefits (see, however, Gaskell *et al.* 2004; Legge and Durant 2010). Second, we see trust as being more relevant for some groups than it is for others. Specifically, it should matter most for the 'Tradeoff' group, and hardly at all for the 'Relaxed', 'Sceptical', and 'Indifferent' groups. In Figure 12.1, the arrow drawn from the trust box into the risk–benefit calculus to support for biotechnologies signifies this interaction effect: if benefits and risks are both perceived, then trust matters. Otherwise, we expect that this trust is irrelevant to a citizen's final decision concerning support for a biotechnology.

Data and methods

To begin testing these ideas here, we draw on a recent nationally representative survey of Canadian attitudes toward genomic science and biotechnology. We designed the survey and Angus Reid Strategies administered it in June 2009.[7] The survey was conducted over the internet using a stratified weighted sample based on demographics taken from the most recent Canadian census (2006). The non-weighted sample size is n = 2019. We queried attitudes toward a range of biotechnologies (see Table 12.1), including detailed modules assigned to subsets of the larger sample. Each respondent answered three modules, assigned so that they varied as much as possible in terms of the field (environmental, agricultural, or medical) and type of technology (that is, genetic modification, biomarkers for selective breeding, and genetic-based testing).

The first step in the analysis examines the distribution of the four risk–benefit groups. We use a question that taps general attitudes toward biotechnology, rather than a specific application, and the following indicators are used to measure benefits and risks respectively:

- Biotechnology will improve the quality of life for all Canadians.
- I am confident that any unexpected outcomes from biotechnology can be controlled.

Responses were measured on a standard five-point Agree/Disagree Likert scale with the middle category being 'Neither agree nor disagree'. The measures were re-coded so that higher values represent stronger convictions of perceived benefits and risks respectively. We then examine predictors of perceived benefits and risks in separate multiple linear regression models as well as a full model of support for biotechnology.[8]

Modelling perceptions of benefits and risks

Before turning to the multiple regression models, Table 12.2 presents the distribution of Canadians based on their perceptions of the risks and benefits of biotechnology. As the table shows, over sixty per cent of Canadians do see potential benefits; however, nearly half of those also see potential risks. It appears that the Tradeoff group plays a critical swing role in determining the public's position toward biotechnology. This is consistent with our analysis of European public opinion based on 2005 Eurobarometer survey data (Weldon and Laycock 2009).

Turning now to the multiple regression models for perceived benefits and risks, overall the results fit well with our core hypotheses. The two factors that are traditionally part of the Deficit Model, scientific knowledge and orientation to science, are significant in each regression.[9] Knowledge correlates positively with both benefits and risks, but with a coefficient nearly twice as large on risk. Orientation to science is more straightforward, with an optimistic orientation increasing perception of benefits and decreasing that of risks. It is clear that these factors act on support for biotech through a risk–benefit calculus.

The findings on knowledge are especially important given their relation to the traditional Deficit Model, which predicts that higher levels of knowledge lead to a decreased perception of risks. Interest in the Deficit Model is high because it suggests an easy solution to allay public concerns. However, recent research has suggested that knowledge may increase perceptions of benefits, but simultaneously increase perceptions of risk (Hamlett *et al.* 2008; Gavelin *et al.* 2007). This is precisely our finding. Importantly, while previous studies have drawn this conclusion by examining participants' perceptions of risk and benefits before and after they participated in a deliberative process, our model demonstrates that the correlation exists among the general population. Controls for attentiveness to science and interest in science further confirm the role that knowledge plays in increasing perceptions of risk.

Table 12.2 Typology of orientations toward biotechnologies in Canada

Perceived Benefit	Perceived Risk	
	Yes	No
Yes	Undecided (29%)	Relaxed (34%)
No	Sceptical (22%)	Indifferent (15%)

Source: Angus Reid Survey, 2009.

(Note: typology adapted from Gaskell *et al.* 2004).

Table 12.3 Linear regression models of benefits and risks

Survey Item	Benefits	Risks
Age (3 pt. scale)	−0.145* (.04)	0.049 (.05)
Income (7 pt. scale)	0.028 (.02)	0.078* (.02)
Female	0.019 (.06)	0.230* (.08)
Knowledge (12 pt. scale)	0.024* (.01)	0.040* (.02)
Orientation to Science (4 pt. scale)	0.190* (.04)	−0.150* (.05)
Attentiveness to Science (4 pt.)	0.060 (.04)	−0.010 (.05)
Interest in Science (4 pt. scale)	0.064 (.04)	0.061 (.05)
Moral Risk (8 pt. scale)	−0.199* (.01)	0.238* (.02)
Religious (No/Yes)	0.012 (.06)	−0.268* (.08)
Post-Materialism (3 pt. scale)	−0.079 (.05)	0.045 (.06)
Constant	2.754* (.23)	0.602* (.29)
Weighted N; R^2	792; 0.34	784; 0.24

Source: Angus Reid Survey, June 2009.

Note: * $p < 0.05$. Values are unstandardised coefficients with standard errors in parentheses.

However, it is important to note that, though these effects are significant, their magnitude is not large. For example, on average, an increase from the mean number of correctly answered questions (6.9) to answering all twelve questions correctly would increase the perception of risk by only 0.24 points. Given that perception of risk is measured on a five-point scale, this is a rather minor effect. Though certainly important, knowledge does not appear to be the key determining factor that the traditional Deficit Model predicts. We should thus not be surprised if the simple solutions suggested by Deficit Model advocates do not deliver the results promised.

The other key finding from the models concerns moral risk, which we measured with one question tapping the belief that biotechnology is 'unnatural', and another tapping the belief that it is 'unethical'. Summing the two measures produced a single indicator of perceived moral risk. The indicator is highly significant in each model, which suggests that moral concerns do indeed cloud perceptions of biotechnologies' benefits and risks. That is, all things being equal, someone who sees biotechnology as a moral risk is much more likely to respond that it is not possible to control unexpected outcomes and to reject the idea that it will improve the quality of life for Canadians.[10]

Taken together, our results indicate that affective factors are the key determinants of risk and benefit perceptions, not scientific knowledge. Moral risk and orientation toward science are the strongest predictors in both models, and, notably, their respective effects are inversely correlated and have a similar magnitude. This suggests that, for many citizens, moral risk and orientation toward science act as general decision-making heuristics that guide perceptions of benefits and risks simultaneously. That is, for example, someone positively oriented toward science is just as likely to perceive benefits from biotechnology as she is to perceive a lack of risk.

Modelling support for biotechnology by risk/benefit group

We now move to a series of models that predict support for biotechnology. To better test the theoretical ideas developed in this chapter, we run separate models for each of the four risk/benefit groups. The dependent variable is support for biotechnology generally, which we measure with five questions about support for different types of biotech applications: genetic testing, assisting selective plant breeding, assisting selective animal breeding, genetic modification of plant organisms, and genetic modification of animal organisms. The response options were:

- I approve of these technologies without reservations or serious concerns.
- I approve of these technologies, as long as the usual levels of government regulation are in place.
- I approve of these technologies if they are more tightly regulated by the government.
- I do not approve of these technologies except under special circumstances.
- I do not approve of these technologies under any circumstances.

Values were summed across the five questions, reverse coded so that higher values indicate greater support, and divided by five. The measure runs from zero to four and acts as a general indicator of support for biotechnology. Table 12.4 shows the results for each risk/benefit group.

The results are interesting, as much for the variables that are significant as for those that appear to have no effect on support once controlling for the pattern of perceived benefits and risks. The strongest predictors in the models are the two newly introduced variables measuring trust in institutional actors as sources of information about biotechnology. As hypothesised, these factors are least important for the Relaxed and Sceptical groups, and more important for the Undecided as well as Indifferent groups. The latter two groups do not have a clear position on the risk-benefit scale, perceiving either both or neither. The findings suggest that these citizens are most likely to turn to experts to help them make policy judgments, using trust in regulators and scientists jointly as heuristics or cues in that process.[11]

Trust in scientists is also significant and positive for the Relaxed and Sceptical groups. We interpret this finding slightly differently for each group. For the Relaxed group, we take it to mean that distrust in scientists can lead One to temper enthusiasm for biotechnology, acting as a type of check on an otherwise positive predisposition. For those in the Sceptical group, however, such distrust will consolidate a negative view of these technologies.

Table 12.4 Predicted support for biotechnologies by risk–benefit groupings

	Relaxed	*Undecided*	*Sceptical*	*Indifferent*
Age (3 pt. scale)	−0.14* (.06)	−0.03 (.06)	−0.12 (.07)	0.23 (.13)
Income (7 pt. scale)	−0.00 (.00)	0.03 (.03)	0.01 (.03)	−0.04 (.06)
Female	−0.23** (.08)	−0.25** (.09)	−0.03 (.11)	−0.32 (.19)
Moral Risk (8 pt. scale)	−0.04 (.02)	−0.07** (.02)	−0.16** (.03)	−0.20** (.06)
PM Values (3 pt. scale)	−0.09 (.07)	−0.11 (.07)	−0.04 (.09)	0.27 (.16)
Knowledge (12 pt. scale)	−0.02 (.02)	0.04 (.02)	0.04 (.02)	0.01 (.03)
Attentiveness (4 pt. scale)	0.05 (.05)	0.09 (.06)	0.02 (.07)	0.11 (.13)
Orientation to Science (4 pt.)	0.12 (.06)	0.06 (.06)	0.12 (.07)	0.03 (.15)
Trust in Regulators (5 pt.)	0.07 (.05)	0.15** (.05)	0.04 (.05)	0.30** (.11)
Trust in Scientists (5 pt.)	0.13* (.06)	0.16** (.06)	0.19** (.05)	0.37** (.10)
R – Squared	0.13	0.26	0.41	0.48
Weighted	28 N	212	151	61

Source: Angus Reid Survey, June 2009.

Note: ** $p < 0.01$; * $p < 0.05$. Values are unstandardised coefficients with standard errors in parentheses.

Moral risk is the only other substantive factor that remains significant in this model as well as those in Table 12.3. The effect is in the expected direction and statistically significant at the 0.05 level for the Undecided, Sceptical, and Indifferent groups, and just outside that for the Relaxed group. This supports our hypothesis that perceptions of moral risk act outside the conventional risk–benefit calculation – at least, in the sense that they are significant, even when controlling for risk–benefit calculations. Finally, the three indicators associated with the Deficit Model, scientific knowledge, attentiveness, and orientation toward science, appear to play no role in policy judgments once we control for perceptions of risks and benefits.

Discussion and conclusion

Our results suggest several tentative conclusions. First, we add an important caveat to the Gaskell *et al.* argument that citizens weigh and consider benefits and risks of biotechnologies in consequentially distinctive ways. Our caveat concerns the role that 'moral risk' – a combination of beliefs that biotech is either unnatural or unethical – can play after the risk–benefit calculation stage. We found that while moral risk correlates strongly with risk in general, it has nearly as strong a negative correlation with benefits. While knowledge of science is likely to shift citizens into the 'Tradeoffs' category of decision makers regarding biotechnologies by increasing their perceptions of both benefits and risks, moral risk and orientation to science drive perceptions in opposite directions. That is, strong moral risk concerns are likely to produce 'Sceptical' citizens, while positive orientations to science lead to 'Relaxed' approaches to supporting biotechnologies.

Such scenarios appear to involve an affect heuristic determining high levels of either perceived risks or benefits and simultaneously low levels of the opposite. In each scenario, the inconsequential weight accorded to risks may appear irrational to those that see only careful risk–benefit balancing to be rational. But following the lead of well established theory regarding 'rational voters', it may be more realistic to portray each of the sceptical and relaxed approaches as rational adaptations to complex informational environments, where people often take short cuts to decisions with the aid of heuristics, cues, trust, and larger value or ideological orientations (Lupia, McCubbins and Popkin 2000). If citizens frequently take such short cuts when considering their support for biotechnologies, we need to face this fact.

'Getting to maybe', with its balanced assessment of benefits and risks, may sit more comfortably with the scientific ideal. But given the powerful impact that 'moral risk' factors have on much public judgment concerning biotechnology, it does not seem realistic to expect that bombarding citizens with more crash-course information on the benefits of biotechnology will significantly reduce the number of sceptics. And if this is so, it seems as likely to be true with regard to public support for first generation as for second and third generation biotechnologies (Stewart and McLean 2004). The broader spiritual, ideological, and value orientation factors that shape perceptions of moral risk are not significantly different for publics now than they were in the years of heightened controversy over GMOs. Although there are some reasons for assuming that 'next generation' biotechnologies will be less welcome now than when they are well established and more comprehensible by citizens, we also have good reason to expect, based on past experience, that public support will vary more according to the fields of their application – agricultural, environmental, and medical – than it will vary according to the complexity of the products or the processes that create them. In other words, the assumption that there is a single, universal logic in the public's judgments on such issues seems as likely to be wrong in the present and foreseeable future as it has been in the recent past. It is, of course, possible that some combination of extended food crises, stunning medical genomic research breakthroughs, and resource-saving environmental genomic applications could substantially reduce the proportion of the population that trumps risk–benefit calculations with moral risk concerns. Only continued careful investigation of public opinion will allow our understanding of these complex decision-making dynamics to grow.

Our second major conclusion concerns the logic internal to the conventionally rational 'Tradeoff' perspective on support for biotechnologies. Underlying our interpretation of this perspective is our finding that knowledge increases public perceptions of biotechnologies' benefits, but also those of risks, as suggested in earlier qualitative studies on citizen deliberation about biotechnological innovation (Hamlett *et al.* 2008; Gavelin *et al.* 2007). Testing the causal model presented here, we demonstrate that the correlation exists between knowledge and pre-existing levels of each variable, and that this effect is directly produced by knowledge rather than by particular attitudes it may represent. This finding reinforces the thrust of our comments above, insofar as they further undermine the 'Deficit Model'. Not only are 'Sceptics' very unlikely to be converted by science-guided public education campaigns, but those citizens who do 'get to maybe' seem to be as open to information that emphasises risks as that which stresses benefits of biotechnologies. So long as we accept that there is an irreducible pluralism in the range of minimally informed positions on these matters, this public openness to knowledge about both benefits and risks seems likely to be reinforced even among those for whom balanced 'Tradeoffs' regarding biotechnologies seems the natural approach. Future research into the best way to model the variety of decision-making approaches employed by citizens will thus be crucial to the design of next generation regulatory regimes consistent with citizens' preferences.

Notes

1 Full question wording is available from the authors on request.

2 It is worth noting that Canada has no regulations requiring producers to label GM food products as such.

3 We speculate that the relative hesitation in supporting biomarkers for selectively breeding salmon (to identify salmon that are able to survive at higher water temperatures) is most likely due to many respondents also associating this technology with farmed salmon. Independent of biotechnology, opposition to salmon farming is relatively high in Canada.

4 Forestry industries are important and high profile in much of Canada, and the public – especially in western Canada – cannot help but have heard about the devastating effect that the mountain pine beetle has had on forests.

5 This does not mean that such individuals are not rational; after all, why continue to weigh costs and benefits if no benefits are seen? At best, one could only be indifferent.

6 To simplify the model, we focus just on individual level and substantively causal factors, thus excluding country level factors, such as economic and political development (see Crawley 2007; Kurzer and Cooper 2007), and standard demographic controls including gender, age, education, and income.

7 This survey was conducted as part of the 'Winegen' Study, funded by Genome BC, the University of British Columbia, and Simon Fraser University.

8 Question wording for independent variables not explained in the text is available on request.
9 We do not have an adequate measure for perceived personal control over the use of the technology. This is something we hope to remedy in future studies.
10 While intended to tap perceived safety risks and physical quality of life, respondents may also have interpreted these questions in spiritual terms.
11 We use the term jointly, because further analysis shows that trust in regulators and trust in scientists are only moderately correlated with one another (0.22).

Bibliography

Barnett, J., Cooper, H. and Senior, V. (2007) 'Belief in Public Efficacy, Trust, and Attitudes Toward Modern Genetic Science', *Risk Analysis*, 27(4): 921–33.

Bennett, P. and Calman, K. (1999) *Risk Communication and Public Health*. Oxford: Oxford University Press.

Coyle, F. and Fairweather, J. (2005) 'Space, Time and Nature: Exploring the Public Reception of Biotechnology in New Zealand', *Public Understanding of Science*, 14(2): 143–61.

Crawley, C. (2007) 'Localized Debates of Agricultural Biotechnology in Community Newspapers', *Science Communication*, 28(3): 314–46.

Durant, R. and Legge, J. Jr (2005) 'Public Opinion, Risk Perceptions, and Genetically Modified Food Regulatory Policy: Reassessing the Calculus of Dissent among European Citizens', *European Union Politics*, 6(1): 181–200.

Durant, J., Bauer, M., Gaskell, G., Midden, C., Liakopoulos, M. and Scholten, E. (2000) 'Two Cultures of Public Understanding of Science and Technology in Europe', in *Between Understanding and Trust: The Public, Science and Technology*. Amsterdam: Harwood.

Evans, G. and Durant, J. (1995) 'The Relationship between Knowledge and Attitudes in the Public Understanding of Science in Britain', *Public Understanding of Science*, 4(1): 57–74.

Fairweather, J. and Cook, A. (2005) *New Zealanders and Biotechnology: Attitudes, Perceptions and Affective Reactions*. AERU. Retrieved from www.lincoln.ac.nz/story9430.html.

Fishburn, P. (1974) 'Lexicographic Orders, Utilities and Decision Rules: A Survey', *Management Science*, 20: 1442–71.

Frewer, L. J., Scholderer, J. and Bredahl, L. (2003) 'Communicating about the Risks and Benefits of Genetically Modified Foods: The Mediating Role of Trust', *Risk Analysis*, 23(6): 1117–33.

Gaskell, G. and Jackson, J. (2005) 'A Comparative Analysis of Public Opinion: Canada, the USA and the European Union', *First Impressions: Understanding Public Views on Emerging Technologies*. Retrieved from http://www.biostrategy.gc.ca/.

Gaskell G., Allum, N., Bauer, M. and Durant, J. (1999) 'Worlds Apart? The Reception of Genetically Modified Foods in Europe and the US', *Science*, July 16: 384–87.

Gaskell, G., Allum, N., Wagner, W., Hviid Nelson, T., Jelsoe, E., Kohring, M. and Bauer, M. (2001) 'In the Public Eye: Representations of Biotechnology in Europe', in *Biotechnology 1996–2000: The Years of Controversy*. London: Science Museum Publications (53–79).

Gaskell, G., Allum, N., Wagner, W., Kronberger, N., Torgersen, H., Hampel, J. and Bardes, J. (2004) 'GM Foods and the Misperception of Risk Perception', *Risk Analysis*, 24(1): 185–94.

Gavelin, K., Wilson, R. and Doubleday, R. (2007) *Democratic Technologies: The Final Report of the Nanotechnology Engagement Group (NEG)*. London. Retrieved from www.involve.org.uk/negreport.

Gottweis, H. (2002) 'Gene Therapy and the Public: a Matter of Trust', *Gene Therapy*, 9(11): 667–9.

Grove-White, R., Macnaughten, P., Meyer, P. and Wynne, B. (1997) *Uncertain World: GMOs, Food and Public Attitudes in Britain*. Lancaster: CSEC, Lancaster University.

Hamlett, P., Cobb, M. and Guston, D. (2008) *National Citizens' Technology Forum: Nanotechnology and Human Enhancement*. CNS-ASU. Retrieved from cns.asu.edu/files/report_NCTF-Summary-Report-final-format.pdf.

Irwin, Alan and Wynne, Brian (eds) (1996) *Misunderstanding Science?: The Public Reconstruction of Science and Technology*. Cambridge: Cambridge University Press.

Kurzer, P. and Cooper, A. (2007) 'Consumer Activism, EU Institutions and Global Markets: The Struggle over Biotech Foods', *Journal of Public Policy*, 27(2): 103–28.

Legge, J. S. and Durant, R. F. (2010) 'Public Opinion, Risk Assessment and Biotechnology: Lessons from Attitudes toward Genetically Modified Foods in the European Union', *Review of Policy Research*, 27(1): 59–76.

Lupia, A., McCubbins, M. and Popkin, S. (eds) (2000) *Elements of Reason: Cognition, Choice, and the Bounds of Rationality*. New York: Cambridge University Press.

Poortinga, W. and Pidgeon, N. (2005) 'Trust in Risk Regulation: Cause or Consequence of the Acceptability of GM Food?' *Risk Analysis*, 25(1): 199–209.

Priest, S. H., Bonfadelli, H. and Rusanen, M. (2003) 'The "Trust Gap" Hypothesis: Predicting Support for Biotechnology Across National Cultures as a Function of Trust in Actors', *Risk Analysis*, 23(4): 751–66.

Siegrist, M. (2000) 'The Influence of Trust and Perceptions of Risks and Benefits on the Acceptance of Gene Technology', *Risk Analysis*, 20(2): 195–204.

Sjoberg, L. (2000) 'Factors in Risk Perception', *Risk Analysis*, 20(1): 1–11.

Sjoberg, L. (2005) 'Gene technology in the eyes of the public and experts: Moral opinions, attitudes and risk perception', *Center for Risk Research, Stockholm School of Economics: Working Paper Series in Business Administration*, no. 2004:7.

Slovic, P., Finucane, M., Peters, E. and MacGregor, D. (2007) 'The Affect Heuristic', *European Journal of Operational Research,* (177): 1333–52.

Stewart, P. A. and McLean, W. (2004) 'Fear and Hope over the Third Generation of Agricultural Biotechnology: Analysis of Public Response in the Federal Register', *AgBioForum*, 7(3): 133–41.

Sturgis, P. and Allum, N. (2004) 'Science in Society: Re-Evaluating the Deficit Model of Public Attitudes', *Public Understanding of Science*, 13(1): 55–74.

Sturgis, P., Cooper, H. and Fife-Shaw, C. (2005) 'Attitudes to Biotechnology: Estimating the Opinions of a Better-informed Public', *New Genetics and Society*, 24(1): 31–56.

Traill, W., Jaeger, S., Yee, W., Valli, C., House, L., Lusk, J., Moore, M. and Morrow, J. L. Jr (2004) 'Categories of GM Risk-Benefit Perceptions and Their Antecedents', *AgBioForum*, 7(4): 176–86.

Weldon, S. and Laycock, D. (2009) 'Public Opinion and Biotechnological Innovation', *Policy and Society* 28(4): 315–25.

Yearley, S. (2000) 'Making Systematic Sense of Public Discontents with Expert Knowledge: Two Analytical Approaches and a Case Study', *Public Understanding of Science*, 9(2): 105–22.

Zaller, John R. (1992) *The Nature and Origins of Mass Opinion*. Cambridge: Cambridge University Press.

13 Deriving policy and governance from deliberative events and mini-publics

Michael M. Burgess

Introduction

Public consultation is frequently required in policy development related to science to enhance representation and trustworthiness (Dietrich and Schibeci 2003; Fukuyama and Furger 2007; House of Lords 2000; Walmsley 2009). It is not without its critics. Public consultation is often criticised for tokenism or superficiality; it imposes costs and the recommendations that follow may inhibit trade and liberty. This chapter narrows the scope to the most ambitious of public consultation – deliberative engagement – that is intended to provide informed, warranted, and civic-minded conclusions. Despite the critiques of public consultation, assessments of specific attempts to support deliberative engagement demonstrate that carefully structured deliberations can provide valuable input into policy processes and implementation. Further, the examples suggest that when a diversity of perspectives and interests are well represented, deliberation focuses on components of governance that the participants believe are important for the trustworthy management of uncertainty.

Critiques of public engagement

The popularity of public consultation has been particularly strong in policy processes related to the environment, science, and technology, sometimes characterised as "democratisation" (cf. Hamlett 2003; Sclove 1995, 1999). There is growing interest in the design of deliberative institutions and processes in policy making, planning, and technology assessments (Fung, Olin, Abers 2003; Fung 2003, Gastil and Levine 2005). The objectives behind consulting publics vary widely, including informing publics, assessing initial reactions and understanding attitudes or values, increasing the representation of marginalised persons, providing an opportunity for stakeholders with different interests, and enhancing policy trustworthiness and legitimacy (Fung 2006; Burgess and Tansey 2008). Rowe and Frewer (2005) characterise over 100 types of public consultation, but agree with Arnstein's (1969) observation that few of these approaches give authority to the conclusions arising from the consultation.

There has been considerable critique of public consultations related to science and technology policy. One category of critique has to do with the influence of public consultation. Technological developments may be too well established for consultations to have any real effect (MacNaghten, Kearnes, and Wynne 2005) or presume technocratic solutions and undermine non-expert input (Kerr, Cunningham-Burley, and Tutton 2007). Consultations may be narrowly framed in a way that stimulates public support for contentious technologies (Petersen 2005; Irwin 2001; Goven 2003). Influence on policy may be ignored as 'cultural' or 'moral' concerns (cf. Jackson 2001, Goven 2006). On the other hand, consultations may carry disproportionate political and social power to legitimate social policies (Dodds and Ankeny 2006; Ankeny and Dodds 2008).

Further, there is a tendency to conceptualise 'science' and 'democracy' as distinct domains rather than as mutually supporting social constructions (Reardon 2007). Some critical debate about deliberative democracy has focused on the definition of reason as problematic (Means 2002; Mouffe 2002; Reardon 2007; Fung 2006). Although social science studies of science and technology demonstrate that publics can understand complex issues sufficiently to have meaningful discussions (Kerr *et al.* 1998), many public consultations are not adequately informed (cf. Niemeyer 2004; Rowe and Frewer 2005; and Sunstein 2002).

Public consultations are not typically subjected to extensive evaluation due to monetary and other constraints (Irwin 2001; Einsiedel *et al.* 2001). Many commentators argue that public consultation should be subject to explicit justifications and assessment (Rowe *et al.* 2005). Some (e.g. Einsiedel 2002) have called for evaluations of concepts that are of particular significance to the field of public consultation such as sampling for representativeness (Irwin 2001; Einsiedel *et al.* 2001).

In the area of health care decision making, Daniels and Sabin have proposed 'accountability for reasonableness' as a basis for justifying decisions about health insurance coverage. They claim that since access to health care services leads to differences in opportunities, decisions about what to include and exclude need to be made fairly. They propose five criteria that constitute the reasons for such decisions: public rationales, relevant evidence and principles, process for appeals, and public regulation (1998: 57). They agree that widely representative participation would enhance fairness, but conclude that it is not realistic:

> Consumer participation might improve deliberation about some matters, but it is unlikely that we could ever enlist active enough consumer participation to deliberation about limit setting. ... there is no realistic mechanism for making consumers who participate truly representative of the consumer population as a whole (Daniels and Sabin 1998: 61).

The next section briefly describes a design for informed public deliberation that has been implemented and assessed in response to these critiques and challenges.

Research on deliberative engagement using mini-publics

Decisions related to biotechnology consume large amounts of public funds and shape many aspects of future society without adequate representation of the diversity of interests present in wider society. Electoral, market, and institutional systems are influenced by well-organised interests groups including industry, research institutions, scientific organisations, and non-governmental organisations (NGOs). Most of these groups ultimately appeal to the public interest to make their cases, usually from particular perspectives that often seem reasonable when considered for particular cases. Policy decisions and their implementation reflect decisions about the relative merit of the accounts of public interests on the assumption that the decision makers can act in the public interest without direct public input. Given the critiques summarised above, and the ethical nature of many of these decisions, it seems reasonable to take the advice of Susan Sherwin to the Canadian Biotechnology Advisory Committee, 'when approaching complex policy matters, we should actively seek out moral perspectives that help to identify and explore as many moral dimensions of the problem as possible' (Sherwin 2001).

This section describes a series of deliberative engagements on issues related to biotechnology to demonstrate that it is possible and desirable to seek informed deliberative input of diverse publics into policy (Burgess *et al.* in press). The design of the deliberative event was the result of several graduate seminars and the activities of a research group leading to a small international workshop in 2006. Following the workshop the research team revised the design and received further comment from the invited participants. The design has undergone subsequent revisions as each event was assessed against the critiques described earlier to estimate the quality of the deliberation and the usefulness of the outcomes.

The design and implementation of two deliberative engagements on biobanks in British Columbia (BC), with additional use of the approach at the Mayo Clinic in Rochester, Minnesota, and the Office of Population Health Genomics in Western Australia, occurred from 2006–09. The deliberative event on salmon genomics took place in Vancouver, BC, in November 2008, and the RDX remediation event in April 2010. The design of the first deliberation was produced through graduate seminars, interdisciplinary collaboration, a small international workshop, and months of refinement. Subsequent events were revised based on review of the preceding events and transcripts, input from peer reviewers of submitted papers and at conferences, and through expanding collaborations to include more areas of science and political science. Nevertheless, the overall structure and process generally follows the original design.

The primary objectives of the design were to:

- form a group of participants that reflected the diversity of perspectives;
- to inform the participants of the technical and contextual issues and opinions;

- to avoid stakeholder capture, ensure respectful listening and participation by all participants;
- stimulate critical appraisal of claims of experts, stakeholders, and participants; and
- encourage the development of group decisions about what might be appropriate policy and identify areas of persistent disagreement.

Representation

There is a great deal of disagreement about what constitutes a 'representative' group for public engagement. The approach described here attempts to represent better the diversity of interests in Burgess (2004) rather than to represent a population through a random sample. The intensity of deliberative engagement makes larger groups expensive and difficult to manage. It is inadequate to accomplish a diversity of participants if the deliberation does not provide an opportunity for all perspectives in the group to be considered. Goodin and Dryzek (2006) proposed that 'mini-publics' or small groups reflecting the diversity in a population could be informed and deliberate to produce an enhanced range of views or interests. These perspectives supplement the expert and stakeholder inputs that are more readily available to policymakers and those charged with implementation of institutional arrangements.

The informed deliberation, and ensuring full participation in the development of the convergences and divergences of opinion, would be difficult to achieve on a larger scale (Goodin and Dryzek 2006; but see Ackerman and Fishkin 2004). There is no guarantee that randomly sampled participants will encompass the full range of interests and values relevant to biobanking. Recruiting specifically for diversity of interests increases the probability of a range of participants who can draw on distinct life experiences, values, and styles of reasoning when contributing to both large and small group deliberations. 'Proxies' for diversity include demographic variables (i.e. ethnicity, religion, occupational group) as well as life experiences that can be anticipated to influence perspectives on the issues (e.g. urban or rural residence may influence experience of health systems; military experience may influence attitudes toward RDX pollution). Tradeoffs in potential sampling methods and the ways in which these tradeoffs may affect achievement of key event objectives were considered in detail in an assessment of representation in the first event (Longstaff and Burgess 2010).

Participants in the public deliberations consisted of a random demographically stratified sample of twenty-five residents of BC, with different events ranging from 24–26 participants (Western Australia and the Mayo Clinic managed their own recruitment similarly). The first events used random digit dialling. Later events mailed letters of invitation to 5000 households that were randomly selected by postal code. The participants were randomly drawn from this pool of individuals to fill the demographic stratification. Participants were compensated for their travel expenses and time at a rate of C$ 100 per day for a total of four days, and completed informed consent procedures to participate in the events as research participants (reviewed and approved by each institution's research ethics committee).

Structure of the deliberative process

The initial design and results of the biobank deliberation are described elsewhere, and summarised here (Avard *et al.* 2009; Burgess *et al.* 2008). Implementing the theory of deliberative democracy, it was important to ensure that the process informed participants and that they were supported in attempts to understand the perspectives of others. Once the group was comfortable with the give and take of deliberation, it was also important to develop an environment in which participants demanded and provided 'warrants' for their positions, and were willing to attempt to find a policy or other conclusions that they agreed to be fair (Gastil and Levine 2005). The process was facilitated to draw participants' attention beyond their own interests toward a 'civic mindedness' that could move toward agreement without diluting differences between individuals' interests (cf. Hamlett 2003: 121–2). Combining Chambers' (2003) and Dryzek's (2000) accounts, we attempted to reflect five general objectives of deliberative democracy:

- Augment legitimacy through accountability and participation.
- Encourage a public-spirited perspective on policy issues through cooperation.
- Promote mutual respect between parties through inclusion.
- Enhance quality of decisions and opinions through substantive and informed debates.
- Allow the contestation of (notably dominant) discourses through the public sphere.

The deliberative events met over two weekends with twelve days between the meetings. The rationale was that it was important to avoid premature drawing of consensus due to exhaustion or group-think. This twelve-day break also gave participants an opportunity to reflect, collect further information, talk to others, and return to the second weekend of deliberation with a stronger sense of the issues that were important to them or others. Large and small group discussions were facilitated to draw out the quieter members, remind participants to listen respectfully, and ensure discussion of points that might be lost without facilitation, or in large groups alone.

The first task assigned to the groups was to get interests and issues on the table using the frame of 'hopes and concerns' before critiquing whether they were realistic, which also helped to promote the sense that everyone could contribute. The first two days of deliberation alternated between receipt of information (discussed below), and facilitated small and large group discussions, with audio recording for further analysis. The emphasis on respectful listening and inclusive participation supported participants' sense that the process was legitimate by encouraging full participation and accountability for explaining oneself and trying to understand each other. This helped to develop a public-spirited perspective on policies issues through cooperation on understanding the range of perspectives in the group and from broader stakeholders and experts.

The task for the second weekend shifted from understanding to group decision making about policy advice. For example, the task in the first biobank event was to identify the values that should inform a provincial biobank and the second was to provide advice on five issues of implementation for the BC BioLibrary. Facilitated large and small groups encouraged inclusive participation and detailed reflection on reasons, narratives, and perspectives. Facilitators encouraged critical reflection and the retelling of others' narratives, and the participants considered arguments and identified persistent disagreements as well as issues that could not be resolved without further information. Facilitators were encouraged to identify persistent disagreements as much as convergence or consensus positions to avoid forcing agreement. This supported a public spirited perspective on policies issues through cooperation on identifying both disagreements and agreement. Again, mutual respect between parties through inclusion was reflected in the process, and the quality of decisions and opinions were enhanced through substantive and informed debates that critically appraised but did not exclude less 'reasoned' perspectives.

Informing deliberation

The legitimacy of both the process of deliberation and the outcomes depends on well-informed dialogue, which presents several challenges. The range of content to present to participants is particularly thorny. For example, the topic of biobanks is often raised in connection with genetic testing, but its broadest sense includes data and tissue repositories supporting longitudinal cohort studies on environmental influences on human health. Opposition to some uses of genetic testing has stimulated criticism of biobanks from groups concerned about disability discrimination, racism, and indigenous rights. Using the broad definition threatens to undermine the concerns that some groups have about genetic uses of biobanks, but failing to include cohort studies may lead to conclusions that could unintentionally undermine cohort studies. The information provided as background materials therefore presented a wide contextual approach, including considerations of expense and lost opportunity costs that may result from funding biobanks, salmon genomics, or remediation of RDX. This broad background also provided a basis for critical appraisal of the claims of experts by participants and between the expert and stakeholder presenters.

How information is presented is important due to the different levels of education, experience and learning styles of the participants, and because the presentation of information will shape expectations about appropriate expressions of perspectives and warrants. Charismatic or offensive speakers can also lead to participants taking positions without adequate critical assessment. As mentioned earlier, a significant critique of deliberative engagement is a narrow definition of reason (Means 2002; Mouffe 2002; Reardon 2007; Fung 2006), and concerns that the engagement is not adequately informed (cf. Niemeyer 2004; Rowe and Frewer 2005; and Sunstein 2002).

Information was presented in a wide range of media. Literature reviews and refining through drafting produced an 18–20 page booklet written for a tenth grade level reading that was circulated in advance of the deliberation, and available online. In most of the events, collections of media and peer reviewed literature, available in print or online, supplemented the booklet. Presenters of stakeholder and expert perspectives were encouraged to consider the booklet and take issue with it to encourage an openly critical approach to information. The issue of undue influence of the expert and stakeholder presenters on the deliberation was addressed by having them present early on the first day but not participate in subsequent small and large group discussions. They were, however, available through the organisers to answer questions throughout the period of deliberation. Assessment of the discussions suggests that participants critically engaged with the speakers' presentations in their small groups (MacLean and Burgess 2010), and that most apparent errors of understanding in discussions did not influence the major themes of the discussion (Wilcox 2009).

Several events used a physical model to locate the issues in a social space and illustrate its connections. Biobank models emphasised connections from community, health clinics, industry, university, government labs, and the way biobanks could be used to support research with a variety of possible outcomes that have effects on populations, environmental concerns, and clinical practice (cf. Burgess, O'Doherty, Secko 2008). The salmon genomics model emphasised the connection of salmon genomics to commercial, aboriginal, and recreational fisheries, the cultural and natural environment from local to global, food systems and knowledge development. Some speakers referred to the models to indicate where they were most concerned, and participants were encouraged to move pieces of the model to reconceptualise the problems and responses, or to indicate where they were most concerned.

As mentioned, the twelve-day break was an important feature of the structure. For the biobanks and salmon genomic events, participants were oriented to a private website. During the break a few participants asked questions, had discussions, reported on additional readings and websites. Several participants reported discussions with their families, friends, and co-workers when they returned on the second weekend. All materials presented on the website were also printed and posted on the walls for review by the participants. The twelve-day break enabled many participants to gather more information, whether from friends and family or online, while for others it was just a break from the intense activity of the deliberation.

Outcomes of deliberation

It is difficult to respond resoundingly to critiques that the outcomes of public consultations are manipulated, or fail to do more than reflect the assumptions or goals of the organisers. Outputs must be legitimately and transparently derived from the event. This begins with the selection of the issues or area presented for deliberative engagement to ensure that they are not so intractably ensconced in institutional practice and culture that outcomes are unlikely to make a difference, particularly if outcomes do not support adoption of technology. The deliberations must be assessed to ensure that they were not narrowly framed in a manner that stimulates participant support for contentious technologies. The approach to informing the participants in deliberative events was discussed earlier, but it is also important to assess whether information was critically assessed and played appropriate roles in the deliberation. Similarly, the approach to achieving a widely diverse participant pool has been discussed, but it is also important to assess whether the diversity among the participants was reflected in the range of perspectives effectively considered in the deliberation. Finally, outcomes are the result of processes, and the structure of the deliberative events and facilitation above explain how we get to group decisions about convergence while giving significant prominence to persistent disagreements.

Assessments and analyses of deliberative events have objectives that are in tension with a simple reporting of participants' ratified group decisions and persistent disagreements. For example, the BC biobank deliberation had several types of decisions arising from the event. First, there are the conclusions of the final large group discussion of the event:

- Strong support for biobanks.
- Governing body independent of funders and researchers.
- Standardising procedures for effectiveness.

While these are the most obvious conclusions of the deliberation, defining the precise parameters of what constitutes the 'results' of a public deliberation is itself a non-trivial exercise. Several types of data were used to assess deliberative quality and further policy implications, including audio recordings that were transcribed and coded using qualitative data analysis software; ethnographic field notes taken by observers of the deliberation; flip chart notes made by facilitators and participants themselves; multiple representations made by small groups to the larger group based on the conclusions of their deliberations; three 'small group reports' by the respective small group facilitators, and ratified by the participants; and the large group report, based on a final one and a half hour discussion in the large group, compiled by an observer (M Burgess), and ratified by participants.

It is useful to differentiate between the *deliberative* outputs and the *analytical* outputs of the deliberation. Deliberative outputs are the ratified collective statements made by participants, and persistent disagreements identified in the process of working toward group agreement. Reporting deliberative outputs does not require analysis beyond presenting a comprehensive overview of participants' explicit collective position and persistent disagreements (not to say that this is objective reporting). Ratification of deliberative outputs requires that they are recognisable by participants as their own positions. For example, the 2007 BC biobank deliberation 'deliberative outputs' were based on the three small group reports and the final large group discussion, ratified by participants. It was also possible to differentiate between two levels of consensus, one achieved in discussions in the large group with all participants, and a second consensus within one or two small groups without being discussed in the remaining group(s) or the final large group discussion. These deliberative outputs became part of the data for analysis.

Ethnographic and transcript analysis of the deliberation yielded important insight about the discursive logics utilised by participants (Walmsley 2010). The analytical lens of positioning theory provides a novel understanding of the way in which participants draw on different aspects of their identity to warrant the positions they take during deliberation (O'Doherty and Davidson 2010). These types of *analytical* output are evaluations of the proceedings of the deliberation that develop a deeper understanding of deliberation.

Deliberative events are directed by organisers to address issues that depend on the organisers' perception of what are critical issues. Assessing whether informed deliberation was possible and whether an informed public would support the development of biobanks influenced the design and implementation of the first biobank event. The primary challenge was therefore to present the problems associated with biobanks and facilitation of the discussions to avoid deliberations being 'captured' by one set of vested interests. The deliberations were minimally structured with only two deliberative tasks:

- discuss your hopes and concerns for biobanking and
- design a BC biobank. This deliberation structure supports a strong degree of confidence in the validity of the full consensus on 'in principle' support for biobanks.

In contrast, the second BC biobank deliberation was oriented to provide specific advice on particular issues in the governance of the BC BioLibrary (Watson *et al.* 2009). Based on confidence from the first event which informed that deliberative participants would support the principle of biobanking, the second event focused on the particular choice that organisers of the BC BioLibrary needed to resolve (O'Doherty and Hawkins 2010). The results of this second deliberation are currently being used by the BC BioLibrary to structure institutional protocols and governance on a more fine-grained level.

At the most basic level of analysis, both deliberations demonstrated that it is possible to have informed deliberative engagement on biobanks in which the participants respectfully engage and produce clear articulations of convergences and divergences of group opinion (Burgess, O'Doherty, and Secko 2008). The manner and substance of discussions suggest that the strong support for biobanks that was ratified by the large group did not simply reflect the dominant discourse of presumed health benefits from investment and facilitation of science. In the large group discussions there was consideration of how even scientific research without health benefits was viewed as a worthwhile output of biobanks. Small groups articulated varying levels of trust in the scientific community, leading to large group discussion of the role of industry in biobanks, with both supportive and critical perspectives being expressed. As reported by Walmsley (2009), one small group elected to present their early discussions as a play rather than through a single representative. Their presentation was about mad scientists and whether the actions of unscrupulous or naïve scientists could lead to unethical uses of biobanks and related research. Other groups raised the issue of unanticipated consequences, sometimes by referring to other instances of science leading to unintended consequences (cf. Wilcox 2009). While it is not possible to rule out that the support for biobanks was a reflection of the unexamined faith in health research, it is also clear that there was explicit opportunity to consider and challenge that assumption. Further, the one strong condition imposed by the participants was governance that is independent of funders and researchers (Burgess, O'Doherty, Secko 2008; Secko *et al.* 2009).

Influence of deliberative events

Demonstrating effective and appropriate influence on policy decisions and implementation is notoriously difficult. The wide range of communications from the deliberations further complicates this assessment. A primary objective was merely to demonstrate that it is possible to achieve informed, deliberative advice that reflects a diversity of public perspectives. The success in that area is reflected in peer reviewed articles assessing the events and in invitations to collaborate on subsequent deliberative events. But participants reasonably expect their investment to have some effect, although the expectations may inevitably be inflated given the policy context.

Deliberations were assessed and reported to funders and policymakers who were responsible for funding and regulating biobanks in Canada, utilised by the Mayo Clinic and the BC BioLibrary in establishing their policy and implementation, and served as critical input for the Western Australia Department of Health (2010). The salmon genomics materials were developed as a report and provided to Genome Canada, Genome BC, collaborating genome scientists, and presented at a workshop with Fisheries and Oceans Canada participation. The materials from the remediation of RDX event are still under preparation, but intended for distribution to Environment Canada and National Defense. Analyses have been presented at international conferences and at invited public and government advisory events.

Sometimes the involvement of an expert or stakeholder facilitates uptake or further development. Peter Watson was a speaker at the biobank event who has responsibility for several biobanks and is leading a project to organise access to collections in BC. Watson assured the participants in the first biobank deliberation that their deliberations would influence the governance of the biobanks for which he had responsibility, and he was key in funding the second BC biobank event. Watson explained that the deliberations could influence the standard of practice for biobanks in Canada and beyond. A participant from the deliberation became a public 'representative' on the BC BioLibrary governing body and was a speaker at the subsequent deliberation in 2009 focused on key policy and practice decisions for the BC BioLibrary. The former participant-representative described her experience at the second BC deliberation (2009) and observed the entire deliberation as a means of informing her ability to work as a public representative.

In addition to the reports described above, the research team for the deliberations presented invited and submitted peer reviewed conferences and policy events as well as publications. These included invitations to speak at workshops and conferences for groups with the responsibility to write policy briefs and revise ethics guidelines relating to genetic testing and biobanks. A colleague from the Mayo Clinic in Rochester, Minnesota, observed the event and collaborated to use the deliberative design at Mayo. The Western Australia Office of Population Health Genomics adopted the design for use with stakeholders, and then again with public participants. The outcomes from their process were used to revise the Organisation for Economic Co-Operation and Development (OECD) draft guidelines (OECD 2009) to write their own draft policy on biobanks. Although difficult to track, these extended communications enhanced the possibility of outcomes arising from the deliberative engagement.

Discussion of analyses across deliberations: governing uncertainty

Trustworthy governance of controversial issues in the context of uncertainty and rapid technological and social change was an analytic theme across the various consultations. As participants grasped the technical information and different social perspectives in the information presented and from other participants, they were directed to consider persistent disagreements and find points of convergence. Persistent disagreement identified areas where there were strong value differences, and where uncertainty made it difficult to use values to determine acceptable risk associated with identifiable and achievable benefits. Convergence toward group decisions sometimes characterised what participants would consider a good decision making process, or at least some of the features of decision making that they would trust. These features across the different topics and consultation suggest components of trustworthy governance for biotechnology. It is not so much the content of these recommendations, but that, as participants recognised the complexity of the decisions and diversity of perspectives, they emphasised trustworthy governance.

Biobanks

Participants generally recognised the value of creating biobanks or data and tissue repositories as a basis for facilitating research. While there was persistent disagreement about when new informed consent was required for a research use that had not been anticipated, there was also a general reduction in concern about the ability to withdraw from a biobank (Secko *et al.* 2009). Participants discussed how it seemed necessary to accept that the future use of samples and data could not be specified at the time of donation, but it was important that each use be identified and not outside of reasonable assumptions at the time of donation. Some participants suggested a mission statement for the biobank, and public representation on the governing body that would identify when uses being considered varied too widely from the mission statement and required new consent or public consultation. When there was no material about research ethics review, participants tended to assume that they could serve as a governing body (i.e. in the 2007 BC deliberation). When there was another governing body for a biobank, participants wanted that group to reflect public concerns and participation (i.e. the 2007 Mayo deliberation). When regulators or implementers of biobanks were directly involved in seeking input from the deliberation, participants did not tend to identify any particular governing body, presumably because they believed they already had direct influence (i.e. Western Australia 2008 and BC 2009 deliberations).

Commercialisation, in the sense of the use of biobanks by industry to produce profitable products, was one of the most controversial issues across biobanking groups, with an emphasis in the discussion on not using public funds and personal donations to biobanks to subsidise corporate profit. This is a clear indication that policies related to managing the relationship with for-profit entities is particularly sensitive and will be a challenge for trustworthy governance. Although it is not clear from the consultations what would justify trust, it is clear that it would be unacceptable to fail to disclose and justify such arrangements, returning to the notion of accountability for reasonableness as a component of trustworthy governance. Biobanks run by for-profit commercial ventures were raised in passing in the discussions, but were not the focus of discussion. The sensitivity to use of public biobanks by for-profit companies raises the issue of how the trustworthiness of commercial biobanks would be assessed by informed public deliberations. Even if the deliberation supported participation in biobanks without full knowledge of specific research on the condition of trustworthy governance, concerns about the influence of profit probably restrict extension of this trust to commercial biobanks.

Salmon genomics

Deliberation about sequencing the genome of salmon was relatively abstract, with participants focusing on how the various contexts in which genome information might be used. Of greatest concern was the use of genomic information in the environment and food. On the positive side were desires to see the benefits of improved environmental monitoring and remediation and safe and secure food supply. On the negative side was the sense that profit and the technological imperative could lead to the inevitable use of knowledge produced by sequencing the salmon genome to modify salmon genetically that would end up in food systems and in the environment. While there was no appetite to close down research to sequence the salmon genome, it was clear that this research was not trusted in the absence of a system that can assess, and when necessary prevent, the introduction of genetically modified salmon into food systems, or risk escape into the oceans and rivers.

Although the salmon genomics deliberation did not provide components of what would constitute trustworthy governance of salmon genomics, participants did emphasise that their trust of the science depended on the existence of mechanisms to assess risks and their authority to restrict unsafe or environmentally hazardous practices. The implication is that, based on the information provided and their own experiences, participants did not believe that existing governance relating to food and environmental regulation established the conditions for them to trust that well meaning scientific research would not be applied to profit at the expense of public and environmental risk.

Environmental remediation

The deliberation on the environmental remediation of explosive and neurotoxin RDX (Research Department Explosive) provided a context in which participants could consider what risks they were willing to consider for remediation, including genetically modified organisms (GMO). There was strong support in the group for assessing whether the use of GMOs provided superior results and were safe, with an emphasis on expert-based assessments. There were strong opponents of the use of GMOs in the environment who did not alter their position through the deliberation, but were required by the deliberation to provide reasons for their position. These participants seemed willing to accept the compromise of assessing GMOs on the basis that they did not believe GMOs would ever be found safe.

The agreement to trust assessment of GMOs for efficacy and safety in the remediation discussion is in contrast to the distrust of existing mechanisms in the salmon genomics deliberation. More information about regulatory mechanisms was provided in the remediation deliberation than in the salmon genomics, which might have helped participants assess current practice. But the tendency in the biobanks deliberations to lean on the most obvious governance body presented – whether research ethics boards, biobank boards, regulators or managers – suggests an alternative interpretation. Deliberative participants facing uncertainty that undermines their ability to draw conclusions may tend to defer to the first available technocratic body to make the assessments.

There is still a point to deliberations that defer decision making about policy and implementation to expert and governing bodies. On the one hand, these conclusions may give credibility to the concern that deliberations reinforce technocratic responses and reinforce the expert–lay knowledge divide. On the other hand, it may also mark out a reasonable division of responsibility, with participants wanting the exercise of expert assessments to be trustworthy, and then either specifying conditions of trust or identifying existing bodies that seem worthy of trust. The conclusion that trustworthy governance requires the capacity to assess efficacy and risks associated with technology may seem obvious in retrospect. But this does not undermine the normative importance of seeking informed deliberative input from diverse publics on what decisions should be referred to experts and governing bodies and the conditions that participants think are required for decision making to be trustworthy. Further, participants are also responding to information that supports an understanding of these assessments as dynamic in the sense that there is a rapid change in the available information and technology that alters the conditions of earlier assessments. It may be feasible to expect participants in a deliberation to make specific decisions with current information, but their focus reasonably includes future scenarios and they assess their own incapacity to make decisions for the future under such dynamic conditions.

Conclusion

Deriving public input from deliberative engagement is possible, but requires significant investment, careful planning and facilitation, assessment and further analysis beyond merely citing the group decisions. Deliberative engagement can provide a context in which a representative mini-public critically assess expert and stakeholder information and consider a range of participant perspectives with the goal of accommodating reasonable diversity while moving toward group decisions about public interests where this reflects the considered views across the group.

Understanding the technical nature of efficacy and risk assessments, deliberative groups can rank what they consider important, and identify the challenge of assessing technical information and the dynamic context for subsequent decisions that is characterised by rapid generation of new knowledge and technology. In response to the challenge of involving diverse public perspectives in the assessment of technical information within a dynamic context, deliberative groups struggle to identify which existing administrative bodies can be trusted to make such decisions and to characterise features that make such bodies trustworthy.

Deliberative groups also identify points of persistent disagreement that may require more active management, and are unlikely to be resolved through further civic dialogue or additional data (cf. Warren 2009). These disagreements anticipate some of the most important areas for policy and administration to actively engage and transparently manage, if they are to sustain public trust. Combining suggestions for governance and responsibility for how persistent disagreements are managed helps move governance from being merely trusted until there is a crisis, to reflecting an approach that makes the governance trustworthy in the sense that deliberative participants believe it can manage these conflicts on behalf of the public.

Deliberative engagement of the type described here is expensive and time consuming, so needs to be employed strategically. In particular, these activities may be useful for policy related issues where expert and stakeholder opinion is divided on what is in the public interest but the wider public is unlikely to form informed opinions that will be sustained when the respondents are given new information or encounter stakeholders. In these cases, public opinion surveys are unlikely to engage a wide diversity of perspectives and interests or produce informed and civic-minded conclusions and the investment in deliberative engagement may useful.

Bibliography

Ackerman, B. and Fishkin, J. S. (2004) *Deliberation Day*. New Haven and London: Yale University Press.

Ankeny, R. A. and Dodds, S. (2008) 'Hearing community voices: public engagement in Australian human embryo research policy', *New Genetics and Society*, 27 (3): 217–32.

Arnstein, Sherry R. (1969) 'A ladder of citizen participation', *Journal of the American Institute of Planners*, 35 (July): 216–24.

Avard, Denise, Bucci, Lucie M., Burgess, Michael M., Kaye, Jane, Heeney, Catherine and Cambon-Thomsen, Anne (2009) 'Public health genomics (PHG) and public participation: Points to consider', *Journal of Public Health Policy* 5 (1), Article 7. http://services.bepress.com/jpd/vol5/iss1/art7.

Burgess, M. M. (2004) 'Public consultation in ethics: an experiment in representative ethics', *Journal of Bioethical Inquiry*, 1 (1): 4–13.

Burgess, M. M. and Tansey, J. (2008) 'Democratic deficit and the politics of "informed and inclusive" consultation', In E. Einseidel and R. Parker (eds) *Hindsight and Foresight on Emerging Technologies*. Canada: UBC Press (275–88).

Burgess, M. M., Longstaff, H. and O'Doherty, K. (in press) 'Assessing deliberative design of public input on biobanks', in R. Ankeny and S. Dodds (eds) *Big Picture Bioethics*. London: Routledge.

Burgess, M. M., O'Doherty, K. and Secko, D. M. (2008) 'Biobanking in BC: Enhancing discussions of the future of personalized medicine through deliberative public engagement', *Personalized Medicine*, 5 (3): 285–96.

Chambers, S. (2003) 'Deliberative democratic theory', *Annual Review of Political Science*, 6: 307–26.

Daniels, N. and Sabin, J. (1998) 'The ethics of accountability in managed care reform', *Health Affairs*, 17 (5): 50–64.

Dietrich, Heather and Schibeci, Renato (2003) 'Beyond public perceptions of gene technology: community participation in public policy in Australia', *Public Understanding of Science*, 12: 381–401.

Dodds, S. and Ankeny, R. A. (2006) 'Regulation of hESC research in Australia: promises and pitfalls for deliberative democratic approaches', *Bioethical Inquiry*, 3: 95–107.

Dryzek, John S. (2000) *Deliberative Democracy and Beyond*. Oxford: Oxford University Press.

Einseidel, E. F. (2002) 'Assessing a controversial medical technology: Canadian public consultations on xenotransplantation', *Public Understanding of Science*, 11: 315–31.

Einseidel, E. F., Jelsøe, E. and Breck, T. (2001) 'Publics at the technology table: The consensus conference in Denmark, Canada, and Australia', *Public Understanding of Science*, 10 (1): 83–98.

Fukuyama, F. and Furger, F. (2007) *Beyond Bioethics: A Proposal for Modernizing the Regulation of Human Biotechnologies*. Washington, DC: Paul H. Nitze School of Advanced International Studies.

Fung, Archon (2003) 'Survey article: recipes for public spheres: eight institutional design choices and their consequences', *Journal of Political Philosophy*, 11(3), 1 September: 338–67.

Fung, Archon (2006) 'Varieties of participation in complex governance', *Public Administration Review*, 66 (1): 66–75. http://www3.interscience.wiley.com/journal/118561477/abstract

Fung, Archon, Wright, Erik Olin and Abers, Rebecca (2003) *Deepening Democracy: Institutional Innovations in Empowered Participatory Governance*. London: New York: Verso.

Gastil, John and Levine, Peter (2005) *The Deliberative Democracy Handbook: Strategies for Effective Civic Engagement in the Twenty-First Century*. San Francisco: Jossey-Bass.

Goodin, R. E. and Dryzek, J. S. (2006) 'Deliberative impacts: the macro-political uptake of mini-publics', *Politics & Society*, 34 (2): 219–44.

Goven, J. (2003) 'Deploying the consensus conference in New Zealand: democracy and de- problematization.', *Public Understanding of Science*, 12 (4): 423–40.

Goven, J. (2006) 'Processes of inclusion, cultures of calculation, structures of power: scientific citizenship and the royal commission on genetic modification', *Science Technology Human Values*, 31 (5): 565–98.

Hamlett, P. W. (2003) 'Technology theory and deliberative democracy', *Science, Technology, & Human Values*, 28 (1): 112–40.

House of Lords (2000) 'Science and Technology – Third Report' Science and Technology Select Committee, Third Report. London: HMSO.

Irwin, Alan (2001) 'Constructing the scientific citizen: science and democracy in the biosciences', *Public Understanding of Science*, 10 (1): 1–18.

Jackson, T. (2001) 'Cultural values and management ethics: a 10-nation study', *Human Relations*, 54 (10): 1267–302.

Kerr, A., Cunningham-Burley, S. and Amos, A. (1998) 'The new genetics and health: mobilizing lay expertise', *Public Understanding of Science*, 7: 41–60.

Kerr, A., Cunningham-Burley, S. and Tutton, R. (2007) 'Shifting subject positions', *Social Studies of Science*, 37 (3): 385–411.

Longstaff, H. and Burgess, M. M. (2010) 'Recruiting for representation in public deliberation on the ethics of biobanks', *Public Understanding of Science*, 19 (2): 212–24.

MacLean, S. and Burgess, M. M. (in press) 'In the public interest: assessing expert and stakeholder influence in public deliberation about biobanks', *Public Understanding of Science*, 19 (4): 486–96.

MacNaghten, Phil, Kearnes, Matthew B. and Wynne, Brian (2005) 'Nanotechnology, governance, and public deliberation: what role for the social sciences?', *Science Communication*, 27, no. 2 (December): 268–91.

Means, A. (2002) 'Narrative argumentation: arguing with natives', *Constellations*, 9 (2): 221–45.

Mouffe, C. (2002) 'Politics and passions: the stakes of democracy', in *Centre for the Study of Democracy Perspectives*. London: University of Westminster.

Niemeyer, Simon (2004) 'Deliberation in the wilderness: displacing symbolic politics', *Environmental Politics*, 13 (2): 347–72.

O'Doherty, K. C. and Burgess, M. M. (2009) 'Engaging the public on biobanks: outcomes of the BC biobank deliberation', *Public Health Genomics*, 12 (4): 203–15.

O'Doherty, K. C. and Davidson, H. J. (2010) 'Subject positioning and deliberative democracy: understanding social processes underlying deliberation', *Journal for the Theory of Social Behaviour*, 40 (2): 224–45.

O'Doherty, K. C. and Hawkins, A. (2010) 'Structuring public engagement for effective input in policy development on human tissue biobanking', *Public Health Genomics*, 13 (4): 197–206.

Organisation for Economic Co-Operation and Development (2009) *OECD Guidelines on Human Biobanks and Genetic Research Databases*. http://www.oecd.org/dataoecd/41/47/44054609.pdf.

Petersen, Alan (2005) 'Securing our genetic health: engendering trust in UK Biobank', *Sociology of Health & Illness*, 27, no. 2 (March): 271–92.

Reardon, J. (2007) 'Democratic Mishaps: The Problem of Democratisation in a Time of Biopolitics', *Biosocieties*, 2: 239–56.

Rowe, G. and Frewer, L. J. (2005) 'A Typology of Public Engagement Mechanisms', *Science, Technology, & Human Values*, 30 (2): 251–90.

Rowe, Gene, Horlick-Jones, Tom, Walls, John and Pidgeon, Nick (2005) 'Difficulties in evaluating public engagement initiatives: reflections on an evaluation of the UK GM nation? Public debate about transgenic crops', *Public Understanding of Science*, 14: 331–52.

Sclove, Richard (1995) *Democracy and Technology*. New York: Guilford Press.

Sclove, Richard (1999) *Democratic Politics of Technology: The Missing Half, Using Democratic Criteria in Participatory Technology Decisions*. Amherst, Mass.: The Loka Institute.

Secko, D. M., Preto, N., Niemeyer, S. and Burgess, M. M. (2009) 'Informed Consent in Biobank Research: Fresh Evidence for the Debate', *Social Science & Medicine*, 68 (4): 781–9.

Sherwin, S. (2001) *Toward Setting an Adequate Ethical Framework for Evaluating Biotechnology Policy*. Ottawa: Canadian Biotechnology Advisory Committee.

Sunstein, Cass (2002) 'The Law of Group Polarization', *Journal of Political Philosophy*, 10 (2): 175–95.

Walmsley, H. L. (2009) 'Mad scientists bend the frame of biobank governance in British Columbia', *Journal of Public Deliberation, 5*(1), Article 6. Document9 http://services.bepress.com/jpd/vol5/iss1/art6.

Walmsley, H. L. (2010) 'Biobanking, public consultation, and the discursive logics of deliberation: five lessons from British Columbia', *Public Understanding of Science*, 19 (4): 452–68.

Warren, M. E. (2009) *Two Trust-Based Uses of Mini-Publics in Democracy*. American Political Science Association Meeting Paper. Available at SSRN: http://ssrn.com/abstract=1449781.

Watson, P. H., Wilson-McManus, J. E., Barnes, R. O., Giesz, S. G., Png, A. and Hegele, R. G. (2009) 'Evolutionary concepts in biobanking – the BC BioLibrary', *Journal of Translational Medicine*, 7: 95.

Western Australia Office of Population Health Genomics Public Health Division, Department of Health (2010) *Guidelines for Human Biobanks, Genetic Research Databases and Associated Data*. http://www.genomics.health.wa.gov.au/publications/docs/guidelines_for_human_biobanks.pdf.

Wilcox, E. S. (2009) 'Does "Misinformation" Matter? Exploring the Roles of Technical and Conceptual Inaccuracies', in *A Deliberative Public Engagement on Biobanks*. MA Thesis, Faculty of Graduate Studies, University of British Columbia, Vancouver, BC, Canada.

14 Second generation governance for second generation GM

Christoph Rehmann-Sutter

In science studies of genomic technologies, the issue of how to categorise types and phases of development in biotechnology is far from settled. Some speak of a distinction between first and second generation genomics, some add a third generation, and some substitute 'next generation' for either the third or both the second and third generations of genomic products. In this chapter, the term 'second generation' will be used quite loosely to point to new rounds of innovation and progress in the development of transgenic crops and animals. The first lines focused mainly on reducing chemical input and increasing yields, for instance by introducing genes for pesticide tolerance, pest resistance, or accelerated growth. For the sake of simplicity I will not further distinguish between second and higher generations here. Second generation genetically modified (GM) agricultural organisms include: plants with improved agronomic performance achieved by targeting traits; plants modified to reduce the costs associated with food processing, such as increased oil content or delayed ripening; cis-gene plants modified to reduce their susceptibility to diseases by introducing genes from wild variants of the same species; GM plants that are going to be utilised as producers of drugs or vaccines; and plants producing raw material for industrial applications like starch, fuels, or textiles (Myhr 2005: 44; Stirn 2005: 82; Marshall 2010).

My focus is on governance and how a *second generation* governance could learn from experiences with first generation GM technologies, in order to improve decision-making procedures with the newer technologies. From the perspective of political philosophy, we can distinguish the *factual* power of governance structures, which can be more or less efficient and sustainable, from *legitimate* forms of institutionalisation of procedures of decision making. With regard to the governance of GM technologies, legitimacy is a key issue that needs to be clarified. An element of governance is legitimate, if it is, or can be seen as, justified and acceptable by those who are affected by it (cf. Habermas 1992). Following Fabienne Peter (2010), I see three main sources of political legitimacy: consent, beneficial consequences, and – most important in contemporary thought – public reason and democratic approval (see also Rehmann-Sutter 2009). Legitimacy, among other criteria, implies fairness. And fairness, of course, is an ethical issue. Second generation governance, I therefore assume, should be *ethical governance*. In this chapter I shall explore several aspects that might determine legitimacy of governance in the field of GM food. How can governance regimes and decision-making procedures be discussed and evaluated from an ethical perspective?

I will start by discussing the fact of international regulatory incoherence and widespread distrust in GM technology. Is it a sign of governance failure? How can discordance in attitudes to GM technology and in regulatory regimes be seen from an ethical perspective? In three subsequent theoretical sections on governance, I will discuss the overall ethical objectives of governance, and then suggest how we might understand the subject matter of governance of GM plants or animals for food. Finally, the concept of governance contains at least four different meanings, which will be distinguished. If ethical governance is the question, then the role of ethics within governance approaches is an issue. This will be the topic of the final section.

Discordance

The previous chapters of this book have given evidence of a wide global gap between affirmative and reluctant approaches towards regulating GMOs in food. Within many countries the status of GM is also deeply controversial. In the European Union, none of the GM plants currently authorised are intended for direct consumption. Other countries, like the USA and China, have embraced the technology. As of 2009, 170 million hectares of transgenic crops have been planted, with their value from seeds and licensing revenues climbing to US$ 10.5 billion (Marshall 2010). It goes without saying that this global discordance with regard to the use of GM in the food sector relates to large differences in regulatory structures and decision-making procedures. The EU approach, which in 1999 led to a de facto moratorium on the commercial approval of all GM crops, was inspired by a relatively strong interpretation of the Precautionary Principle (Myhr 2005). In the USA, the Familiarity Principle is used to assess the environmental or health risks of new GMOs. There, safety assessment by the Food and Drug Administration depends on the application of the concept of 'Substantial Equivalence' of GM products to their non-GM counterparts (Myhr 2005). While substantial equivalence might finally be established for a GM salmon that grows twice as fast as wild Atlantic salmon (on the AquaBounty Technologies application see Marris 2010, Egan 2011 – the application was pending at the time of writing), procedural equivalence clearly does not hold. Other considerations are brought forward by consumer and environmental groups, like the labelling of products as 'transgene' (normally absent in the USA), or the risks that the sterile GM salmon could mutate to become fertile and escape to the wild and mate with wild salmon.

This heterogeneous situation, both inter- and intra-nationally, provides ample reasons for complaint by those who are thwarted in their interests, or by others whose concerns are not sufficiently heard. However, from an ethical point of view, it is not clear whether discordance in such a new field of biotechnology is undesirable. There are different ways to assess GM plants and animals ethically (see Mepham 2008, chapters 10 and 11), and there is no consensus as to whether, at the end of the day, GM is ethically good or bad. One can therefore posit the value of comparing experiences with different governance models. By evaluating the differences, the approaches could improve by learning from each other.

From an ethical point of view, it would be problematic, without a general consensus about the ethical acceptability of GM technology in the food sector, to start from the assumption that either the restrictive or the permissive approach to GM for food is right and the other is defective. Ethics can only start by suspending given moral judgments. Ethics is not moralism but a systematic and circumspect *reflection* on moral arguments, regarding how they are perceived, structured, processed, and which key concerns need to be considered when looking for the good solution. The ethical stance is not a particular moral standpoint within a moral controversy, but a search for clarification by interdisciplinary reflective research, as for instance in ELSA (ethical, legal, and social aspects) programmes. This must be also a neutral standpoint. (But this does not mean that ethicists should not come to a moral conclusion and engage themselves politically, for instance for the protection of the rights of the most vulnerable groups.) Only by temporally postponing the moral conclusion can we gain space for a fair and open *ethical* discussion.

However, in a comparative perspective, certain systematic questions need to be posed. If a governance regime is to be fair, how can fairness be understood in this context, and how can it be implemented? The fairness question can be put more concretely: what is fairness in the producer–consumer relationship? How should the concept of 'governance' be interpreted from the point of view of fairness? Fairness (and justice/injustice/exploitation) questions also concern the selection of the targets of technological development, which, in turn, contain the question: who should benefit and who is required to accept burdens or disadvantages? Different parties need to be considered: current breeders, farmers, the food industries, consumers, but also the present and future environment, farm animals, and the plant and animal life of the regions concerned (the 'biota', per Mepham 2008: 50).

First generation governance has been concerned with risk and safety issues, but has run into problems. It could not establish trust in the new technology, either in general or as it is actually handled. The 2010 Eurobarometer shows that since 2005, trust in GM food in the European Union has even decreased, reaching its lowest level of just five per cent of convinced support and another eighteen per cent who said that they 'tend to agree' (Gaskell *et al.* 2010: 37; the numbers in 2005 were six per cent and twenty-one per cent respectively). Public antagonism towards this technology has grown. A new international grassroots organisation of eco-activists, the European Field Liberation Movement, even aims at civil disobedience by destroying GM plantations.[1]

In a Guardian Science Blog, immediately after the European Commission's decision to allow BASF's GM potato Amflora for purely industrial use, Irish plant scientist Eoin Lettice argued that public antagonism regarding agro-biotech has intimately to do with the question of fairness, 'Consumers genuinely do not see the worth of GM products, which is why there is a need to move beyond crops that confer benefits to industry and growers alone, towards second generation GM that produces added health and nutritional benefits for consumers' (Lettice 2010). The selection of the beneficiaries of the new technology seems to be biased. As long as this remains the case, many consumers, at least in the European contexts, will remain unconvinced by GM food and prefer other kinds of food. Without consumers, there is no market; without a market, there are no incentives for an innovative biotechnology.

The answer, however, will not be this straightforward. Lettice quotes from a survey conducted in 2005 by Ireland's Agriculture and Food Development Authority, which showed that forty-two per cent of Irish consumers would purchase hypothetical GM-produced yoghurt, and even forty-four per cent would purchase a hypothetical GM-produced dairy spread if they had anti-cancer properties (Lettice 2010). The task for policymakers would then just be to create incentives for technological development to go in this direction and to exploit these potentially available markets, instead of trying the impossible.

I do not believe that this analysis reaches deep enough. Public reluctance to consume GM products does not exist simply because they provide no direct consumer benefit. Consumers do not just pursue their own interests. Consumers are at the same time citizens. They have more fundamental worries, which are rooted in broader concerns that can be supported by arguments and are related to issues of justification. These worries cannot be addressed by pointing to some potential niches in a future market. They refer to long-term environmental risks, and to concrete issues of social justice and sustainability. They have to do with the way the industrial–technological complex is seen to react to issues like global warming, the risks of nuclear technology, the loss of biodiversity, and a global increase in poverty. They are tied to a critique of globalization and heavy industrialisation in agriculture, where more and more short-term profits have been made by larger and larger multinational companies, where power is unequally distributed between corporate industry and farmers (not to speak of consumers), and where political authorities react rather than act (Levidow and Carr 2010, Potthast *et al.* 2005). These *ethical* concerns inhabit a different argumentative register, and require different kinds of political answers.

The ends of governance

First generation regulation has been preoccupied with risk assessment, risk control, the freedom of the market, coexistence, and the protection of consumer choice according to individual preferences (for the EU case see Kurzer and Skogstad, and Johansson, this volume). Many regulators seem to have shared the assumption that science and technology, in combination with the system of intellectual property rights, a free market, and clear safety procedures, automatically lead to innovations that are ultimately good for society.

I believe that we have good reasons to question this 'automatism assumption'. First, there are examples of effects of technological developments, which have emerged from similar (though perhaps not as strict) oversight systems, but are bad for society.

The destruction of global biodiversity for instance: pollution or global warming will be detrimental to a sustainable functioning of the ecological human–nature relationships (Miller and Spoolman 2009). Second, the automatism assumption rests on a form of technological determinism. According to this view, innovation is first generated within a free domain of science and engineering, combining ingenuity, feasibility, and the anticipation of social needs. Then it goes out into society. Technological determinism, however, has been empirically refuted by science and technology studies (MacKenzie/Wajcman 1999). Arie Rip and Peter Kemp have called it the 'cannonball view' of technological development (Rip and Kemp 1998: 387). In fact, processes of innovation are more complex, more contingent, and feature influences going forward and backward. Multiple social actors are pushing and pulling in many directions (Johnson 2007). 'Society' does more in shaping the processes of innovation. Therefore, the role of society in technology is intrinsically more ambitious than the automatism assumption presupposes. Technology does not evolve independently from society but as sociotechnical systems, which 'consist of artefacts together with social practices, social arrangements, social relationships, and systems of knowledge' (Johnson 2007: 26).

If the process of technological innovation is inherently social, it is not automatic. And if state authorities do more than just filter out unsafe applications, guarantee intellectual property regimes, and ensure the freedom of the market, we need to re-assess basic questions about the ends of social and political life within this special context. This relates to key questions in political philosophy (Arendt 1993; Kymlicka 2002): What is the meaning of the political? What is the role of the state? How is state authority constituted, and what are the limits of its legitimacy?

This task cannot be resolved in a few paragraphs. Different conceptions of the role and the limits of the state have been brought into the discussion, and we are all aware of the plurality of different understandings of how state authority should be constituted. Here I can only point out which answers I see as the most promising.

Role and limits of the state

Following Martha Nussbaum's reading of Aristotle's political philosophy, I would explain the overall aim of the state in the following way. The task of the state is to secure the material, institutional, and pedagogical conditions for all citizens, which provide access to a good human life. They should have the capabilities to perform all those practices or functions, which belong to a good life (Nussbaum 1990). This formulation captures the best interests of all. The term 'citizen' should not, however (quite in contrast to Aristotle's own views about slaves, women, and workers), be used in a discriminatory way that privileges legal citizens over non-citizens who live in the same place. In a cosmopolitan society, the role of the state extends also to the good life of those who happen to live in the territory, and of those outside whose life is affected by the actions of the state. The citizen versus non-citizen distinction is particularly relevant in agro-biotech because today, in many countries, plantation fieldwork is done by migrant workers or other underprivileged people under bad conditions that make them particularly vulnerable.

The state also has limits, which are implied in this explanation of the task of the state: individuals should have the freedom to decide which practices or functions they want to perform. The state should not organise the good life but provide the conditions under which human capabilities can be exercised. Biotechnological progress, understood as a sociotechnical reality, should contribute to the good life of all. Therefore, it should not only be new or fancy, nor just be beneficial for some, but should serve a common public interest. This is an ambitious requirement that means, among other things, that governance is not fully (and fairly) developed if it simply mediates the interests of the different sectors of society.

Constitution of state authority

Political participation by all citizens was a concern for Aristotle, as Nussbaum (1990) has explained. To make choices and to participate in the community's collective decisions ought to be a basic human capability for everybody in society. This participatory role of citizens in the constitution of state authority needs to be further clarified. I follow Jürgen Habermas' concept of a deliberative democracy. Deliberative democracy goes far beyond the distribution of ballots and the collection of votes. It requires the institutionalisation of appropriate forms of substantial communication. 'Everything depends on the conditions of communication and on the procedures, which provide a legitimising force to the institutionalised forms of opinion-making and decision-making' (Habermas 1996: 285, my translation). The procedures of deliberative politics are the normative core of a theory of democracy as a legitimate way of establishing and directing state authority. Deliberation means an open exchange of views, concerns, arguments, proposals, and so on, in a political public sphere, where no power should prevail except that of the better argument. Such a 'political public sphere' is, as Habermas explains it, to be seen as 'an arena for the perception, identification and for the treatment of problems with social relevance' (ibid.: 291; cf. Rehmann-Sutter 2009).

I have a slightly broader view about what can constitute a 'good argument' than Habermas seems to hold. My view is less rationalistic and would also include narrations, experiences and, in appropriate ways, the language of emotions. Some people are less able to formulate their concerns in clear words than others, yet their concerns should also be heard and respected (Young 2000).

In the governance of biotechnology, as we shall discuss below, some parts are more democratic and others less so. Within a government–governance continuum (Lyall 2007), democratic states are explicitly bound to democratic procedural rules. This gives them the capability to establish democratic legitimacy for their decisions concerning biotechnology, a capability lacking in systems without a democratic constitution, such as markets or multinational companies.

These two aims of governance can be discussed in more detail, in order to set more concrete criteria for the ethical evaluation of governance in the sector of GM food.

The producer–consumer relationship

An important direct implication stems from the theoretical reconstruction of food GM biotechnology as sociotechnical systems, in which the artefacts shape social practices, become involved in social arrangements, shape social relationships, and are embedded in systems of knowledge (Johnson 2007). The implication is that we can see more clearly *what* governance of technology is actually shaping. If biotechnology is *social practice*, not just a GMO put into an environment where it could prove beneficial or dangerous, the *locus* in society where food biotechnology is socially realised, is the producer–consumer relationship. Food is, by definition, a product to be consumed (i.e. incorporated and metabolised) by humans. Food can therefore only be realised in a relationship between people who produce and people who buy, cook, and eat or drink the product. The socio-technical system of GM food is constituted by producer–consumer relationships. This means that a new way of producing food not only changes the plants, animals, and the product itself, but essentially the producer–consumer relationship.

In the simplest form of home gardening, the producers and the consumers may be closely related. But the social arrangement can be more complex, as in the case of a group where one part of the group takes responsibility for the garden. In the case of traditional farming, the producer–consumer relationship is realised as an interaction between farmers, traders, and clients. Subsequently, more agents have been introduced into this relationship (cf. Goody 1982; Mazoyer and Roudart 2006): landowners, workers, breeders, engineers, food processing factories, transport, wholesalers, retailers, and so on. In the case of industrial agriculture of the present time, the producer–consumer relationship has grown more complex still. It encompasses global markets, labour, multinational companies, financial speculation, global transport, and scientific research institutes – but it still exists; the consumer still buys a product and eats or drinks it. In the act of *incorporation*, which is both a bodily and a moral act, the consumer participates in an extended producer–consumer relationship. Some consumers, however, are more aware of the moral implications of consumption than others.

The growing complexity of this relationship and its partial lack of transparency explains why consumers have a growing demand not only to know what they eat – that is, how the product is materially composed – but also how it is produced, where, by whom, and under what circumstances. The relationship of consuming relates consumers to the persons and circumstances of production in a particular way. Therefore, the procedural aspects and the social conditions of production matter to the consumer. The current trends towards fair trade, organic farming, vegetarianism, and veganism are motivated by these consumers' desire to eat and drink *meaningful* products, not just the cheapest, nor even necessarily the most savoury ones. The joy of eating is more than simply finding a dish delicious. The joy of eating comprises the practice of buying and cooking, serving and the companionship of eating, and it is affected also by the more extended social and ecological reality of the producer–consumer relationship, which materialises in the meal on the plate. Hence, for these consumers, the meaning of the product exceeds its material composition. It includes the effects of farming and transport on the affected ecosystems, the effects on other humans and on animals, even on plants. The socially aware food consumer, I believe, is not just a *homo economicus* who is looking for the best deal, but also a *moral actor* who is conscious of her or his role within different possible producer–consumer relationships.

The producer–consumer relationship, in most cases, is not a close relationship between people who know each other personally. It is in many ways indirect and mediated. The larger the distance becomes between producer and consumer, and the more partners participate, the more important trust becomes. Trust is important because the consumer is existentially dependent on the producer. Without food the consumer could not live. The producer–consumer relationship is, therefore, also a dependency relationship.

In a dependency relationship, trust relies, however, not only on the subjective components of good relationships ('I believe that you care about my best interests'), but also on their objective components, which are realised in the structures of society and in governance. The question of trust on this level is rather 'Can I have confidence that the producers' own interests do not clash with my own best interests as a consumer?'[2] According to the Aristotelian approach to good governance, the aim of political structures is the good life of the citizens. Therefore, the harmonisation of interests in the producer–consumer relationship is a key topic of ethical governance. Trust in a dependency relationship relies also on transparency. The essential decisions about the use of GM in the food chain, therefore, need to be taken in the public sphere. Otherwise, participation would not be possible and decisions would lack public legitimacy.

The forms of governance of GM food shape the producer–consumer relationship in a very important way. Concretely, governance inserts controls, norms, structures, and constraints into this relationship and thereby affects the level of trust. This is not a new phenomenon. Jack Goody gave a lively account of the challenges of this relationship during industrialisation. Complaints against the adulteration of food are even older, 'as old as the sale of foodstuffs itself. In Athens protests about the quality of wine led to the appointment of inspectors to control its quality. In Rome wines from Gaul were already accused of adulteration, and local bakers were said to add "white earth" to their bread' (Goody 1982: 171). The more divorced from primary production an urban society is, the more important and the more challenged is consumer trust. The regulation of the producer–consumer relationship is the task of good governance. We could say governance of GM food becomes a social reality exactly by *organising* the producer–consumer relationships. To me this seems to be the key ethical aspect in GM food governance. The meaning of food encompasses the structure of the producer–consumer relationship, including the dimensions of trust and all its social, political, and ecological ramifications.

Let us discuss one concrete example to see how this frame works. Is it necessary to label GM products visibly in the supermarket as 'transgene'? In a situation where all consumers would be indifferent to the question of GM, labelling would not increase the level of trust in the producer–consumer relationship. But in our present situation, where a large segment of consumers strictly refuse a product *because* it is produced by using GM plants or animals, labelling is a requirement of fairness. Non-labelling would be unfair because it makes GM and non-GM products indistinguishable for consumers. Therefore, from the perspective of the consumers, *all* products would become potential GM products. Those who refuse GM would then be forced by the producers to accept products that are all potentially made with GM technology. Consumers depend on food and therefore, without labelling, they have no other choice than to consume *despite* their distrust in the way it was potentially produced. I cannot see how such an arrangement could be morally justified vis-à-vis this diverse community of consumers. Even without going into depth concerning possible moral arguments and counterarguments, it is quite obvious that non-labelling impairs trust in this relationship.

Similarly, the ethical dimensions of the other foci and motivations for GM regulation – food safety, ecological sustainability, non-interference with other forms of producing, sustainable coexistence of different farming strategies, the effects on biodiversity, ecological risks, eco-toxicity, the use of chemicals, and so forth – can also be further clarified within the frame of the producer–consumer relationship. Within this frame, very importantly, the consumers become major actors. The consumers are not just on the receiving end of an implementation and production chain. It is a common topic in much of the literature on human 'foodways' throughout the ages that people's eating habits somehow express who they are (Kraemer 2007: 1). They constitute *identity*. Roland Barthes for instance said that food is 'a system of communication... food signifies' (quoted in Kraemer 2007: 1). When we *read* food within the producer–consumer relationship as a system of communication and significations, this provides us with an interpretative key for understanding the implications of GM plants and animals in consumers' identities.

The relationship, therefore, must be read from both sides. In their dependency, their need for trust, and in their identity, consumers are discovered as one of the ethical subjects in the relationship. The consumers in contemporary foodways are agents who are in many ways related and dependent. We see an active form of dependency here, or a passive form of activity. Agency here becomes, as Soran Reader (2007) has suggested for other cases, 'patiency'. In both ways, consumers need to be recognised as ethical subjects.

Now we need to clarify the term 'governance' and how it can be used in a political-theoretical argument.

Four concepts of governance

As it has emerged in the political and sociological literature of the last two decades, the term 'governance' refers to the increased role of non-government actors in policy making (Lyall *et al.* 2009: 3). This phenomenon is particularly impressive in the context of biotechnology, where the key developments cannot be driven by state commands, but emerge from opportunities within the unforeseeable and sociologically very complex interactions between new scientific discoveries, the development of research tools, biotech companies' business strategies, venture capital investors, and the market. The state has *a* steering role (by its research funding programmes, subsidies, safety regulations, approval mechanisms for new food products, etc.), but it cannot autocratically plan biotechnological progress. This relationship between governments and non-government actors can be realised (and theorised) in many different ways. The state can, for instance, be more or less hierarchical, attempting to restrict non-state actors by imposing limits and prohibitions, or it can try to 'herd' non-state actors with positive incentives in a direction that state agencies, legislatures, and/or executives believe is in the public interest. This interactive process involving state and non-state actors is governance. The result of governance is a particular shape of society, a particular way of integrating biotechnology into practices, a particular set of producer–consumer relationships.

Evidently, it is not only the state that must be looked at when thinking about good or bad, efficient or inefficient, sustainable or unsustainable, robust or more fragile ways of shaping and integrating society. What matters, rather, is the whole system of these interactions, and the continuum between governments and non-government actors (Lyall 2007). The state is a special actor whose nature differs from other actors in many ways. It assumes supreme power in society by controlling the police, the military, legislation, and the system of jurisdiction. This represents a key public steering and mediating role with regard to private interests. The state authorities are the most important providers of policies by which the state plays its part in governing complex actor networks in society. The role of the state in this process of policy making can be understood and realised in different ways, as is amply demonstrated by both the history of thought about the questions of political justice, and the diversity of historical examples of states. The current discussion mostly starts from the assumption of a democratic state, which is built on a constitution, formulates the basic rights of all individuals, and acts on the principle of law.

Governance of biotechnology, in contrast to top-down regulation of biotechnology, comprises the interactive process of shaping bio-society and its biotechnology-related practices by the state as well as by non-state actors. This contrast then, as the literature shows, can be understood in different ways. I see four different meanings of the term governance:

1   a strategy of steering,
2   the process of de-centralising power,
3   a quality of steering practices, or
4   the totality of steering activities across the different parts of society.[3]

The *strategic* (1) meaning of the term governance has been very clearly explained by Catherine Lyall and Joyce Tait (cf. also 6 Perri 2005):

> 'Governance' is seen as implying a move away from the previous *government* approach (a top-down legislative approach which attempts to regulate the behaviour of people and institutions in quite detailed and compartmentalised

ways) to *governance* (which attempts to set the parameters of the system within which people and institutions behave so that self-regulation achieves the desired outcomes), or put more simply, the replacement of traditional 'power over' with contextual 'powers to' (2005: 3).

Governance in this sense is a practical approach to steering: a particular way of *doing* (planning, implementing, etc.) the steering job. What is specific to governance is defined in contrast to government. Government is steering by top-down legislation, regulating the behaviour of people and institutions. Governance is steering by parameter setting within a system where people and institutions behave in self-determined ways. Efficient governance sets the parameters prudently: self-regulation by people and institutions should achieve desired outcomes. Good (or ethical) governance would set the parameters well in an ethical sense: self-regulation by people and institutions should achieve outcomes that harmonise with the public interest and improve the lives of everybody by making lives more liveable and the institutions more just.

Looking at *governance as decentralisation of power* (2), in reference to a broad literature, Theo Papaioannou explains the demands that are suggested by the strategy of governance:

> ... the political state must share its power with non-political agents, forming partnerships and networks with the private sector. ... non-political actors are gradually allowed to coordinate themselves. ... Heterarchy is considered to be the only alternative process of legitimation to hierarchy of government (2009: 22).

The term 'heterarchy' (introduced by B. Jessop 1998) contains a certain enthusiasm. It implies that hierarchies are less apt, less efficient, less adequate in modern techno-societies, or even less just, because they centralise power instead of sharing or distributing responsibility in open decision-making procedures. The enthusiasm for society-centric approaches, for networks and self-regulation is a priori leaning towards deregulation, decentralisation and redistribution of power within the government–governance continuum.

However, questions about legitimacy must be raised. Is heterarchisation serving society as a whole? Or is it mainly advancing the particular interests of a few, promoting their particular concepts of the good? There are the interests of the researchers, the biotech industry, agro and food business, farmers, and consumers. There are the interests of those living in the present and those living in the future. These are particular demands. However, there are also 'aggregate demands' like justice, which cannot be fulfilled by serving the individuals who have their shares in the power networks. Papaioannou (2009: 23) rephrases the classical question, which has been raised by the different approaches to political justice, in the context of fair distribution of opportunities and risks in genomics, 'Such [aggregate] demands are political and normative while governance, as conceived by its society-centric proponents, is more problem solving than aggregating and overlooks certain political and normative issues'. Seen from this perspective, there are limits to governance (Papaioannou 2009: 45). Heterarchical governance might not be able to deal with specific genomics-related concerns. Papaioannou's study (on human related genomics) concludes that 'only a combination of integrated or "joined-up" policies of a sovereign power and democratic engagement of citizens can overcome the limits that the challenge of genomics creates to governance.' (ibid. 2009: 46) This needs to include a government perspective as well as a cosmopolitan perspective.

In the food sector this argument seems to be especially pertinent. Here, the risk that individual agents, and at least some strategies of deregulation and heterarchisation, are structurally unable to meet aggregate demands, like long-term sustainability and distributive justice, is quite evident, and has been discussed critically ever since Garrett Hardin's famous parable of a *tragedy of the commons*. However, a nuanced view needs to be taken. Elinor Ostrom's research has shown that, under certain circumstances, bottom-up approaches *can* overcome the tragedy of the commons and establish a sustainable local order, which indeed serves the aggregate interests of justice (Ostrom 1990). Certain factors make it more probable, but other impeding factors make it less probable. The question, therefore, might be less likely to be answered by finding the right trade-off between state and governance (the 'right limits' to governance) than by studying and improving concrete rules and systems of decision making, and infusing them with the capability of reflective decision making in a long-term perspective. Among the reasons that endorse heterarchisation are ethically relevant ones, such as the argument that shared responsibility is more sensitive and powerful. We have, therefore, a true dilemma, and not just a trade-off between the state (representing justice) and society (representing private interests). As Joyce Tait has put it:

> The governance-based approach was promoted in a spirit of optimism as a means to achieve more democratic and more robust political processes and decisions, distributing power more equitably across societal groups. However, in many of the cases described here [in the collection by Lyall et al. 2009], the outcome has been greater complexity, which has acted to create a different sort of democratic deficit – a shift in the locus of the power base without corresponding improvement in the responsibility with which that power is exercised (2009, xiv).

If a more equitable distribution of power across societal groups, away from the central state to civil society, is not accompanied by the same kind of scrutiny with regard to equity, social justice, and participation of those affected, the process is prone to justified criticism. This argument is particularly serious because the legitimacy of the redistribution of power (in heterarchisation) has been explained in terms of equity and participation.

The *quality of steering practices* (3) is an evaluative meaning of the term governance. If we ask about the best forms of including governance (meaning (2) above) in the government–governance continuum, we are interested in the moral criteria for evaluating such strategies, or we are looking for best practices of governance. The term governance can be used in this way. Governance, in contrast to government, can refer to the *mode, style* and *manner* of governing, the way governing is *done*, the *approaches* taken, the *strategies* coined and selected, the *mentality* of governing. This sense of governance relates to Foucault's term '*gouvernementalité*' (see Burchell *et al.* 1991).

This question, as we have just said, is particularly relevant in a system where governments and non-government actors interact. Governance (meaning (2)) as heterarchisation is just one *way* of governing, one way of using governance as a strategy (meaning (1)). The strategy selected needs to be scrutinised with regard to qualitative issues of procedural justice, such as transparency and fair inclusion.

Governance as *the totality of steering activities across the different parts of society* (4) is a purely descriptive meaning of the term. We can use 'governance' to describe the processes of steering, shaping, and regulating social systems and institutions on multiple levels by a multitude of participants: those who are empirically relevant. In contrast to 'government', governance in this sense refers to the form of interaction between steering authorities, as diverse as they may be, integrating top-down regulation by the state and legitimate engagement of other players in society, including bottom-up dynamics. This relates to Weiss' formulation, which refers to the Tokyo Institute of Technology definition:

> The concept of governance refers to the complex set of values, norms, processes and institutions by which society manages its development and resolves conflict, formally and informally. It involves the state, but also the civil society (economic and social actors, community-based institutions and unstructured groups, the media, and so on) at the local, national, regional and global levels (2000: 797 f.).

James Rosenau has also introduced the term in this sense. Governance 'encompasses the activities of governments, but it also includes the many other channels through which "commands" flow in the form of goals framed, directives issued, and policies pursued' (1995: 14).

Governance is a suitable term to refer to all processes that define which game will be played and which rules need to be respected in playing this game. There is governance from inside a collaboration (by standard operation procedures, hierarchies, instructions by steering committees, etc.) and governance from outside collaboration (by law, government authorities, patient groups, citizen participation, and so on).[4]

The role of (bio)ethics in governance

Bioethics has a double role in the fourth meaning of governance. It finds itself to be part of the strategy taken. It provides arguments, improves decisions made and the procedures used to make them, and thereby adds legitimacy even where it is critical of final decisions. But, at the same time, bioethicists can (and should) critically reflect on their own role under ethical considerations. There must be an *ethics* of bioethics. Of course, a self-reflective bioethics, which improves the decision-making procedures and their outcomes, will also contribute to legitimacy. But this is no criticism of bioethics per se, quite the contrary; this is exactly what bioethics should do. When asking how much governance or how much government is best, questioning the role of the state in a mixed strategy, as well as the roles of all the other actors, it raises a set of highly relevant political–ethical issues regarding governance.

The role of bioethics in governance has been criticised by Helga Nowotny and Giuseppe Testa (2011). I explain some of their critical points because they give me the opportunity to elaborate on my own view of the role of bioethics.[5]

Instead of just neutrally describing the political processes contributing to governance, Nowotny and Testa provide a critical assessment. They focus on the role of bioethics as a provider of legitimacy in biotechnology. The role of bioethics as 'an instrument of governance' is, however, itself a question. It is, in my view, a contraction of a larger challenge that should be put high on bioethics' own self-reflective agenda in the future. What is the proper role of bioethics in assessing the social implications of emerging biotechnologies? My assumption is that bioethics should be essentially a *reflective* form of serious engagement with the management of interdependencies. It is part of social reflexivity, which cannot blindly do useful things in the interest of powerful actors.

Nowotny and Testa observe seven tendencies inherent in models of governance, which all have implications for self-reflective bioethics. I want to sketch out how I see these implications:

1   Governance as a concept has a tendency to eclipse real power relations and to assume that the world has become nearly flat. If this is true, bioethics should remind those involved in governance processes of the reality of these power relationships, and of the power that is intrinsic in, or becomes possible with, certain uses of artefacts. Bioethics can then be a sensitive detector of undercover bio-power,[6] and positively contribute to governance in a critical way.

2   Governance assumes that a diversity of actors enhances the quality of decision making, regardless of their influence. If this happens, there is a danger that factual differences in authority are ignored. It is simply not the case that everyone can exert the same amount of direct or indirect pressure. In human relations, there are frequently groups or persons who are silenced, actually or potentially oppressed, and exploited. Bioethics needs to give them a voice that appears in the steering processes, a voice that they could perhaps not find by themselves.

3   In the framework of governance, decision-making processes become iterative, a constantly fluctuating negotiation and adjustment, through which compromises are found and interests are attuned to each other. This feature of governance works in favour of bioethics, which defends the possibility of improving and adapting regulatory solutions whenever essential aspects of the situation change.

4 If governments, NGOs, and industry (or other players), with their very different internal decision-making rules that are not all democratic, merge and build joint networks of governance, there is an evident danger of losing democratic legitimacy. If this happens, bioethics must have an active interest in creating transparency, in order to make shared decisions publicly accountable. This again is a critical role that bioethics should assume, a role that is perhaps against the short-term interests of some of the stakeholders, but which contributes to the long-term interests of all.

5 Arguing and bargaining become more important than the counting of votes. This related development can also be seen positively, because it makes governance more hospitable to ethics. Bioethics is more about arguments than about voting. It can also have a role in making the discussions that finally lead to a vote, less prejudiced, more informed and more comprehensive.

6 The ethical discourse and the struggle for 'binding ethical foundations' plays an unexpectedly large role in negotiating processes, which today lack a generally obligatory normative framework. Therefore, 'wide spaces of interpretation' are created to find compromises between seemingly incompatible positions and valuations. Here I see two dangers: vagueness and formalism. A compromise may be stabilised by reference to a very abstract framework of moral principles (abstract enough that all the parties can identify), but remain watery and hollow. Bioethics needs to defend its role of being inconvenient, by listening to precise and full stories across cultural contexts, by searching for thick descriptions, by pointing out differences, and by identifying common concerns and lines of conflict, even if it is against the interest of those aiming for smooth outcomes.

7 Governance creates a public realm, to which all members of civil society are admitted. This is good indeed, but Nowotny and Testa add a caution, 'due to the lack of common values, it tends to be restricted to the smallest common denominator'. But is this necessarily so? Before accepting this tendency as unalterable, the question of the role of cross-cultural, universal values in bioethics should be discussed. Novotny and Testa seem to follow the view that common values should work like a universal covering law, where solutions for particular contexts can be derived. In political theory, this view has been contested with strong arguments, very prominently for instance by Michael Walzer (2007; cf. Rehmann-Sutter, in press), who does not suggest particularism, but an alternative form of universalism, 'reiterative universalism', which assumes that certain moral experiences of justice and injustice can be repeated even in very different socio-cultural frameworks. What is universal is an entitlement to take these experiences seriously. This may lead to a different way of communicating values and norms across cultural divides, which is more appreciative of the other, less principle based, and more open to the full stories and to common *concerns* that are derived from experiences, not from theory alone.

Conclusions: ethical governance

While food GM biotechnology progresses into a new generation, policy and regulation also need to be considered from an ethical point of view. We can learn from comparative policy analyses of the impacts of first generation developments in this area. First generation governance has been mainly focused on health and risk assessment, the avoidance of harm to the biosphere, coexistence and freedom of choice, and on establishing a system where investments into innovative science can become profitable. The GM products that have been developed under first generation governance have been largely opportunity driven and served the interests of companies and producers. Assessments of social needs and demands were not driving the decision making, but rather assessments of market opportunities, or in some cases first world scientists' ideas that a certain product will help the poor and undernourished.[7] From an ethical perspective on progress and on governance as politics, we could point out elements of a vision for second generation governance that could encourage new rounds of decision making.

Whereas first generation governance has favoured *opportunity* driven innovation, legitimate second generation governance should favour *need* driven development. This could be an interpretation both of the Habermasian appeal for publicly accountable and justified decision making, and of the Aristotelian/ Nussbaumian defence of the state as an instrument for a good life for all. It will make developers and growers more independent in their problem solving activities. It will recognise consumers as active moral evaluators and ethical subjects in a producer–consumer relationship. It will encourage consumers to develop their food ethics. It will see the biotechnological product in its social reality: shaping the producer–consumer relationship and the meaning of food. It will include interdisciplinary bioethics and ELSA research to enhance a society's capabilities to develop collective agency and social reflectivity.

Innovation cannot just be market driven, even though it still needs to work with the rules of the market. Stronger participation of farmers, consumers, environmentalists, biotech experts, ELSA experts, and companies should contribute to making agriculture globally more sustainable, safe, reliable, and beneficial to societies. Their participation can lead to the development of GM products that are truly beneficial. What counts as a benefit will depend on their contribution to the wellbeing of consumers (added health and nutritional benefits), to the thriving of ecosystems when compared to other forms of sustainable agriculture, and to social justice. Such a GM biotechnology could perhaps positively contribute to the foodways-related identity of farmers and of consumers.[8]

Notes

1 In Switzerland, where a general moratorium is in force until 2013 that only allows research on GM plants, eco-activists have vandalised all test fields at Reckenholz and Pully. The current security costs equal those of the research itself. Researchers complain that if this situation does not change, scientific fieldwork with GM plants will come to a halt in Switzerland (Hofmann 2011).

2 Russell Hardin's approach to trust is relevant here. He analyses trust as an 'encapsulated interest'. With trust, he demonstrates that the interests align and cooperation is made possible. This is an interesting approach because we can see why conflicts of interests are like the antidote to trust (and cooperativity). 'The trusted party has incentive to be trustworthy, incentive that is grounded in the value of maintaining the relationship into the future. That is, I trust you because your interest encapsulates mine, which is to say that you have an interest in fulfilling my trust' (Hardin 2002: 3).

3 I do not claim that this list is exhaustive or unique. See an extensive literature accessible through Kiaer (2009).

4 This last paragraph is inspired by the explanation of governance of the BIONET Expert Group (2010: 31 f.). Governance in the fourth sense of the term has been used there to discuss ethical and procedural issues in Chinese–European research collaborations.

5 The comments on Nowotny and Testa have previously been published as Rehmann-Sutter (2010).

6 I use the term 'bio-power' here in a Foucauldian way as the use of power in society by means of biological characteristics of humans, such as being dependent on particular kinds of food.

7 The 'Golden Rice' example is discussed in Mepham 2008: 285–9.

8 I thank Rowena Joy Smith and Daisy Laforce for reviewing the English and for helpful suggestions. I am grateful to the editors for their encouragement and for critical comments.

Bibliography

6, Perri (2005) 'The Governance of Technology', in Lyall, Catherine and Tait, Joyce (eds) *New Modes of Governance. Developing an Integrated Policy Approach to Science, Technology, Risk and the Environment*. Aldershot: Ashgate (19–44).

Arendt, Hannah (1993) *Was ist Politik? Aus dem Nachlass herausgegeben von Ursula Ludz*. München: Piper.

BIONET Expert Group (2010) *Recommendations on best practice in the ethical governance of Sino-European biological and biomedical research collaborations*. London: LSE (www.bionet-china.org).

Burchell, Graham, Gordon, Colin and Miller, Peter (eds) (1991) *The Foucault Effect. Studies in Governmentality*. Chicago: University of Chicago Press.

Egan, Timothy (2011) 'Frankenfish Phobia', *New York Times*, 17 March.

Gaskell, George, Stares, Sally, Allansdottir, Agnes, Allum, Nich, Castro, Paula, Esmer, Yilmaz, Fischler, Claude, Jackson, Jonathan, Kronberger, Nicole, Hampel, Jürgen, Mejlgaard, Niels, Quintanilha, Alex, Rammer, Andu, Revuelta, Gemma, Stoneman, Paul, Torgersen, Helge and Wagner, Wolfgang (2010) *Europeans and Biotechnology in 2010. Winds of Change?* A report to the European Commission's Directorate-General of Research. Brussels (http://ec.europa.eu/research/research-eu).

Goody, Jack (1982) *Cooking, Cuisine and Class. A Study in Comparative Sociology*. Cambridge: Cambridge University Press.

Habermas, Jürgen (1992) *Faktizität und Geltung*. Frankfurt am Main: Suhrkamp.

Habermas, Jürgen (1996) *Die Einbeziehung des Anderen. Studien zur politischen Theorie*. Frankfurt am Main: Suhrkamp.

Hardin, Russell (2002) *Trust and Trustworthiness*. New York: Russell Sage Foundation.

Hofmann, Markus (2011) 'Zu teuer und zu aufwendig. Wissenschaftler verlangen ein forschungsfreundliches Gentechnikrecht', *Neue Zürcher Zeitung*, 28.

Jessop, B. (1998) 'The Rise of Governance and the Risks of Failure: the Case of Economic Development', *International Social Science Journal*, 155: 29–45.

Johnson, Deborah G. (2007) 'Ethics and Technology "in den Making": An Essay on the Challenge of Nanoethics', *Nanoethics*, 1: 21–30.

Kaier, L. P. (2009) *Meta-analysis of variety mixtures. Cereal crop interactions inferred from field trial data*. PhD Thesis, University of Copenhagen.

Kraemer, David (2007) *Jewish Eating and Identity Through the Ages*. New York: Routledge.

Kymlicka, Will (2002) *Contemporary Political Philosophy. An Introduction*. Oxford: Oxford University Press, Second edn.

Lettice, Eoin (2010) 'We need GM plants that benefit consumers and not just farmers', *The Guardian*, 9 March.

Levidow, Les and Carr, Susan (2010) *GM Food on Trial: Testing European Democracy*. London: Routledge.

Lyall, Catherine (2007) 'Governing genomics: New governance tools for new technologies?', *Technology Analysis and Strategic Management*, 19: 365–82.

Lyall, Catherine and Tait, Joyce (2005) 'Shifting Policy Debates and the Implications of Governance', in Lyall, Catherine and Tait, Joyce (eds) *New Modes of Governance. Developing an Integrated Policy Approach to Science, Technology, Risk and the Environment*. Aldershot: Ashgate (3–17).

Lyall, Catherine, Papaioannou, Theo and Smith, James (2009) 'The Challenge of Policy Making for the New Life Sciences', Lyall, Catherine, Papaioannou, Theo and Smith, James (eds) *The Limits to Governance. The Challenge of Policy making for the New Life-Sciences*. Farnham: Ashgate (1–17).

MacKenzie, Donald and Wajcman, Judy (1999) 'Introduction', in *The Social Shaping of Technology*. 2nd ed. Buckingham: Open University Press (3–27).

Marris, Emma (2010) 'Transgenic fish go large. Approval expected for genetically modified salmon', *Scientific American*, September: 14.

Marshall, Andrew (2010) '2nd-generation GM traits progress', *Nature Biotechnology*, 28: 306.

Mazoyer, Marcel and Roudart, Laurence (2006) *A History of World Agriculture. From the Neolithic to the Present Crisis*. London: Earthscan.

Mepham, Ben (2008) *Bioethics. An Introduction for the Biosciences*. Oxford: Oxford University Press, 2nd edn.

Miller, G. Tyler and Spoolman, Scott E. (2009) *Living in the Environment. Concepts, Connections and Solutions*. Brooks-Cole: Cengage Learning, 16th edn.

Myhr, Anne Ingeborg (2005) 'Genetically modified organisms: The need for precautionary motivated science', in Thomas Potthast, Christoph Baumgartner and Eve-Marie Engels (eds) *Die richtigen Maße für die Nahrung. Biotechnologie, Landwirtschaft und Lebensmittel in ethischer Perspektive*. Tübingen: Francke (41–72).

Nowotny, Helga and Testa, Giuseppe (2011) *Naked Genes. Reinventing the Human in the Molecular Age*. Cambridge (MA): MIT Press.

Nussbaum, Martha C. (1990) 'Aristotelian Social Democracy', in R. B. Douglas, G. Mara and H. Richardson (eds) *Liberalism and the Good*. New York: Routledge (203–52).

Ostrom, Elinor (1990) *Governing the Commons: The Evolution of Institutions for Collective Action*. Cambridge: Cambridge University Press.

Papaioannou, T. (2009) 'Human gene patents and the question of liberal morality', *Genomics, Society and Policy*, Issue 4, Article 3.

Peter, Fabienne (2010) 'Political Legitimacy' in Edward N. Zalta (ed.) *The Stanford Encyclopedia of Philosophy*, (http://plato.stanford.edu/entries/legitimacy).

Potthast, Thomas, Baumgartner, Christoph and Engels, Eve-Marie (eds) (2005) *Die richtigen Maße für die Nahrung. Biotechnologie, Landwirtschaft und Lebensmittel in ethischer Perspektive*. Tübingen: Francke.

Reader, Soran (2007) 'The other side of agency', *Philosophy*, 82: 579–604.

Rehmann-Sutter, Christoph (2009) 'Bioethical Decisions and the Public Sphere: A Cross-Cultural Perspective' in Brigitte Nerlich, Richard Elliott and Brendon Larson (eds) *Communicating Biological Sciences. Ethical and Metaphorical Dimensions.* Aldershot: Ashgate, (75–91).

Rehmann-Sutter, Christoph (2010) 'The ambivalent role of bioethics: Two diagnoses', *BioSocieties*, 5: 399–402.

Rehmann-Sutter, Christoph (in press) 'Der universalistische Fehlschluss in der transnationalen Bioethik', *Verhandlungen zur Geschichte und Theorie der Biologie*.

Rip, Arie and Kemp, Peter (1998) 'Technological Change', in Steve Rayner, Elizabeth L. Malone (eds) *Human Choice and Climate Change. Vol. 2. Resources and Technology.* Columbus, Ohio: Battelle Press (327–99).

Rosenau, James N. (1995) 'Governance in the twenty-first century', *Global Governance*, 1: 13–43.

Stirn, Susanne (2005) 'Genetically modified foods – risk assessment, regulation, and labelling', in Thomas Potthast, Christoph Baumgartner, Eve-Marie Engels (eds) *Die richtigen Maße für die Nahrung. Biotechnologie, Landwirtschaft und Lebensmittel in ethischer Perspektive*. Tübingen: Francke (73–85).

Tait, Joyce (2009) 'Foreword', in Lyall, Catherine, Papaioannou, Theo and Smith, James (eds) *The Limits to Governance. The Challenge of Policy making for the New Life-Sciences.* Farnham: Ashgate (xiii–xiv).

Walzer, Michael (2007) *Thinking Politically: Essays in Political Theory*. Binghampton: Vail Ballou Press.

Weiss, Thomas G. (2000) 'Governance, good governance and global governance: Conceptual and actual challenges', *Third World Quarterly*, 21: 795–814.

Young, Iris Marion (2000) *Inclusion and Democracy*. Oxford: Oxford University Press.

Index

abortion: German debate 170; US debate
174–5
agricultural biotechnology: areas of
activity and regulatory issues 57, 65–7;
Canadian public opinion 205–6; case
study of salmon genomics deliberation
using mini-publics 231–2; Chinese
governance 111–15, 122–6; controversy
15, 20; East Asia 111–26; Europe 15–31;
Japanese governance 111–14, 115–19,
125–6; Knowledge-Based Bioeconomy
(KBBE) 15, 26, 28, 29; Korean
governance 111–13, 123, 125; national
regulatory regime variations 57–60;
research in China 120–1; safe making
17–20; safety trials 20–2; scientific risk
assessment 43–4, 130; survey of EU, UK,
US and Canadian policy actors 189–201;
United States 17; US federal regulation
103–4; *see also* genetically modified
organisms (GMOs); GM crops; GM
food; GM products; plant biotechnology
agriculture: links with energy 28–9
animal feed: EU propose imports of
trace amounts of GM in place of zero
tolerance 137–8; GM product content
129, 134; risk assessment of GM
imports 137–8
Argentina: WTO complaint against EU
moratorium on GM crops 95, 104–7, 131
ART *see* assisted reproductive technologies
Asilomar conferences 75–6
assisted reproductive technologies
(ART) 164; Belgian regulation 174,
175–6; Canadian regulation 174,
176; French regulation 173; German
regulation 170–1; Italian regulation 174;
Netherlands regulation 171–2; policy
development 165–6; regulation 166–77;

Swiss regulation 171; UK regulation
171, 172–3; US regulation 174–5
Austria: anti-GMO 132, 133, 134–5

Belgium: assisted reproductive technologies
(ART) regulation 174, 175–6
Berg Letter 60, 61
biobanks: deliberative engagement case
study 230–1
bioethics: role in governance 249–52
biomedical technology 164–7; Canadian
public opinion 205–6; *see also* assisted
reproductive technologies (ART)
biopiracy concept 18
biosafety *see* safety; safety regulations
biotech crops *see* GM crops
biotechnical inventions 18
biotechnology: areas of activity and
regulatory issues 57, 65–7; Canadian
public opinion 203–17; case study of
biobank deliberation using mini-publics
230–1; EU regulatory evolution 61–2,
63–4, 97–101; EU regulatory regime
59–60; governance for second generation
GM 3–4, 7, 237–52; national regulatory
regime variations 57–60; politics of
GMO regulation in the European Union
73–89, 129; regulatory evolution 60–4;
research and development in the EU
18–19, 132; standards and guidelines
61; survey of EU, UK, US and Canadian
policy actors 189–201; US regulatory
evolution 60–1, 63–4, 101–4; US
regulatory regime 58–9; *see also*
agricultural biotechnology; genetically
modified organisms (GMOs); plant
biotechnology
BSE (bovine spongiform encephalopathy)
crisis 114

Canada: assisted reproductive technologies (ART) regulation 174, 176; Chinese imports 121; public opinion on biotechnology innovation 203–17; survey of biotechnology policy actors 189–201; trade effect of China's GMO regulation 120, 121; WTO complaint against EU moratorium on GM crops 95, 104–7, 131

Cartagena Protocol on Biosafety 35, 59; China 120, 123; Japan 116; ratified by East Asia 111

chemicals: regulation in China 155

China: agricultural imports 120; chemicals regulation 155; Consumer Council 124; GMO governance 111–15, 122–6; GMO imports 120; GMO production 121–2; importer of GMOs 121; importer of grains and soy 120; labelling 119–20, 122–3; nanotechnology 150–1; nanotechnology development 151–3; nanotechnology regulations 153–61; nanotechnology safety regulations 156–61; research on agricultural biotechnology 120–1; safety regulations 121, 122

citizens *see* civil society

civil society: campaigns 7–8, 20, 97; Canadian public opinion on biotechnology innovation 203–17; Chinese governance of GMOs 114–15, 124–5; Chinese opinion on GMOs 123–4; deliberative engagement in policy development 220–33; democratic representation 186–8; East Asian campaigns 113, 125; EU citizens low support for GM food 135; GM food issues 239–40, 241–5; Japanese lobby on agricultural biotechnology 116, 117, 118–19

Codex Alimentarius 59

consumers: Chinese governance of GMOs 114–15, 124–5; Chinese opinion on GMOs 123–4; East Asian campaign groups 113; GM food issues 239–40, 241–5; Japanese lobby on agricultural biotechnology 116, 117, 118–19; Novel Food Regulation 1997 79, 80–1, 84–5

Convention on Biodiversity (CBD) 97; Biosafety Protocol 105–6

culture: response to GMOs 113

democratic representation 186–8

democratic sovereignty: agricultural biotechnology debate 22–3

EFSA *see* European Food Safety Authority

embryo related research 164; Belgian regulation 175–6; Netherlands regulation 171–2; policy development 165–6; regulation 166–77; US regulation 174–5; *see also* embryonic stem cell research (ESCR)

embryonic stem cell research (ESCR) 164, 165, 166; Canadian regulation 176; France 173; German regulation 170; Swiss regulation 171; UK regulation 173; US regulation 174–5

energy: links with plants 28–9

environment: GM crops eco-efficiency versus agro-industrial hazards 23–5; GMO regulatory policy in the EU 100–1; nanotechnology safety regulations in China 156–61; risk assessments criticised 136

environmental remediation: deliberative engagement case study 232–3

ESCR *see* embryonic stem cell research

ethics: GM food 239; moral risk concerns on biotechnology 213–14, 215, 216; Novel Food Regulation 1997 79, 84–5; role of bioethics in governance 249–52

European Commission: Deliberate Release Directive 90/220/EEC 19, 61, 77–9, 84–5, 98; Directive 2001/18EC 80–1, 85–6

European Field Liberation Movement 239

European Food Safety Authority (EFSA) 62, 129; criticism 136; risk assessment of GM animal feed imports 137–8; scientific risk assessment 130, 135–6

European Union: adaptation of US safety model 17–20; agricultural biotechnology 15–31; biotechnology regulatory evolution 61–2, 63–4, 97–101; biotechnology regulatory regime 59–60; democratic representation 187, 188–9; field trials 131; GM crop approvals 134; GM labelling 23, 62; GM product licensing 130–1; GMO policy compared with US 95–107, 128; GMO politics 73–89, 129; GMO traceability rules 128; moratorium on use of GMOs 1999 62, 79–80, 85, 86, 99; national differences on support for GM food 133, 134–5; Novel Food Regulation 1997 21, 79, 80–1, 84–5; plant biotechnology risk regulation 128–42; proposal for imports of trace amounts of GM in place of zero tolerance 137–8; proposal

to devolve regulatory authority over
GM crop cultivation to member states
138–40, 142; REACH regulation 154–5;
retail market exclude GM food 140;
safety regulations 21, 36; survey of
biotechnology policy actors 189–201;
WTO complaint against moratorium on
GM crops 95, 97, 104–7, 131

field tests: European Union 131; Japan
116, 119; *see also* trials
food security 27
France: assisted reproductive technologies
(ART) regulation 173

genetic modification (GM) technologies
34; China 121; *see also* agricultural
technology; GM crops; GM food; GM
products; plant biotechnology
genetically modified organisms (GMOs)
34; Chinese governance 111–15, 122–6;
Chinese production 121–2; coexistence
controversy 81–3, 86–7; coexistence
politics 87–8; cultural response 113;
deliberative engagement case study
232–3; East Asia 111–26; EU Directives
61–2, 98–9; EU moratorium 1999
62, 79–80, 85, 86, 99; EU regulatory
evolution 61–2, 63–4, 97–101; EU
traceability rules 128; EU versus US
policy compared 95–107, 128; imports
by China 121; Indian production 122;
Japanese governance 111–14, 115–19,
125–6; Korean governance 111–13,
123, 125; Novel Food Regulation 1997
21, 79, 80–1, 84–5; politics of biotech
regulation in the European Union
73–89, 129; precautionary principle
80–1, 99–100; standards 61; US federal
regulation 103–4; US production 101,
128; US regulatory evolution 60–1,
63–4, 101–4; *see also* GM crops; GM
products; plant biotechnology
genomics *see* assisted reproductive
technologies (ART); biotechnology;
genetically modified organisms
(GMOs); GM crops; GM food; GM
products; human genetics
Germany: abortion debate 170; assisted
reproductive technologies (ART)
regulation 170–1
globalisation: agricultural biotechnology
debate, 22–3

GM crops 34, 41; China imports 121; EU
approvals 134; EU proposal to devolve
regulatory authority to member states
138–40, 142; national regulatory regime
variations 57–60; regulatory burdens
38–9, 40; WTO complaint against
EU moratorium 95, 97, 104–7; *see
also* genetically modified organisms
(GMOs); GM food; GM products; plant
biotechnology
GM food: Canadian public opinion 205, 206;
consumer issues 239–40, 241–5; ethics
239; EU citizens low support for 135;
GM-free labelling 140–1; governance
discord 238–42; labelling 19, 23, 62, 111,
115–17, 118, 119, 122–3, 244–5; national
differences in the EU on support for 133,
134–5; retail market exclusion in the EU
140; risk assessment 41–3, 128–9, 135–6;
safety assessment 21; *see also* genetically
modified organisms (GMOs); GM crops;
GM products
GM products: animal feed containing 129,
134, 137–8; blockages 15, 26; Convention
on Biodiversity Biosafety Protocol 105–6;
eco-efficiency versus agro-industrial
hazards 23–5; globalisation versus
democratic sovereignty 22–3; governance
for second generation 3–4, 7, 237–52;
licensing in the EU 130–1; safety claims
versus precaution 20–3, 34; *see also*
genetically modified organisms (GMOs);
GM crops; GM food
governance: GMOs in China 111–15,
122–6; GMOs in East Asia 111–26;
GMOs in Japan 111–14, 115–19, 125–6;
GMOs in Korea 111–13, 123, 125; role
of bioethics 249–52; second generation
GM technologies 3–4, 7, 237–52; *see
also* regulation
GM technologies *see* genetic modification
(GM) technologies
GMOs *see* genetically modified organisms
(GMOs)
Greenpeace: role in Chinese GMO
governance 124

hazards *see* risks
health and safety: nanotechnology safety
regulations in China 156–61
human genetics: survey of EU, UK, US and
Canadian policy actors 189–201; *see also*
assisted reproductive technologies (ART)

in vitro fertilisation (IVF) 164, 165, 166; Belgium 175; France 173; Italy 174; United Kingdom 172; *see also* assisted reproductive technologies (ART)
India: GMO production 122
intellectual property rights 62
intracytoplasmic sperm injection (ICSI) 164
Italy: assisted reproductive technologies (ART) regulation 174

Japan: effect of GMO regulations on economy 113; field tests 116, 119; GMO governance 111–14, 115–19, 125–6; labelling 115–17, 118

Knowledge-Based Bioeconomy (KBBE) 15, 26; concept 26–7, 28, 29
Korea: GMO governance 111–13, 123, 125

labelling: China 119–20, 122–3; European Union 23, 62; GM food 19, 23, 62, 111, 115–17, 118, 119, 122–3, 244–5; GM-free 140–1; Japan 115–17, 118; mandatory in East Asia 111, 125
licensing: GM products in the EU 130–1

'mad cow' (BSE) crisis 114
medical rDNA 40–1

nanotechnology 149–50; China 150–1; development in China 151–3; regulations in China 153–61; safety regulations in China 156–61
The Netherlands: assisted reproductive technologies (ART) regulation 171–2
NGOs (non-governmental organizations) 7–8, 19, 20; Chinese GMO governance 124; Japanese campaign on agricultural biotechnology 116, 117–19

OECD (Organisation for Economic Co-operation and Development) standards 59, 61
organic crops 88

patent rights 18, 62
pharmaceutical industries 40–1
plant biotechnology: EU risk regulation 128–42; *see also* agricultural biotechnology; genetically modified organisms (GMOs); GM crops; GM food; GM products
plants: eco-efficient 27; links with energy 28–9; *see also* GM crops

politics: assisted reproductive technologies (ART) regulation 170–6, 177; China's pursuit of nanotechnology 157–9; GMO regulation in China 122–4; GMO regulation in the European Union 73–89, 129; GMO regulation in north-east Asia 125; Japanese Diet committee on GMO labelling 118
preimplantation genetic diagnosis (PGD) 164, 165, 166; *see also* assisted reproductive technologies (ART)
public consultation: deliberative engagement in policy development 220–33
public opinion *see* civil society

rDNA (recombinant DNA) technologies 34; Asilomar conferences 75–6; development 35; pharmaceutical industries 40–1; risks 40–1; standards and guidelines 61
regulation: assisted reproductive technologies (ART) 166–77; Belgium on assisted reproductive technologies (ART) 174, 175–6; Canadian assisted reproductive technologies (ART) 174, 176; chemicals in China 155; convergence 57; country differences 56–7, 63–4; definition 50; EU plant biotechnology risk 128–42; EU versus US GMO policy compared 95–107, 128; evolution of biotechnology regulation 60–4; France on assisted reproductive technologies (ART) 173; Germany on assisted reproductive technologies (ART) 170–1; Italy on assisted reproductive technologies (ART) 174; lifecycles 51–6, 63–4; national biotechnology regimes 57–60; Netherlands on assisted reproductive technologies (ART) 171–2; Switzerland on assisted reproductive technologies (ART) 171; UK on assisted reproductive technologies (ART) 171, 172–3; United States on assisted reproductive technologies (ART) 174–5; variation of regimes 50–6; *see also* governance; safety regulations
reproductive cloning (RC) 166, 172, 175, 176
research and development: agro-food-forestry-biotech sectors 26–7, 28; China on agricultural biotechnology 120–1; EU policy on biotechnology 18–19, 132; science and nanotechnology in China 150–1
retailers: GM food excluded 140

risk: EU plant biotechnology regulation
128–42; GM crops eco-efficiency versus
agro-industrial hazards 24–5; rDNA
technologies 40–1
risk assessment: GM animal feed imports
137–8; GM food 41–3, 128–9, 135–6;
scientific procedures 43–4, 130

safety: agricultural biotechnology 17–20;
Cartagena Protocol on Biosafety 35, 59,
111, 116, 120, 123; GM risk assessment
41–3, 128–9, 135–6, 137–8; trials 20–2
safety regulations 21, 34–41; China
121, 122; European Union 21, 36; EU
REACH regulation 154–5; international
trade implications 38; language problems
36–7; nanotechnology in China 156–61;
problems 35–9; science-based 39–41;
scientific flaws 38–9
salmon genomics: deliberative engagement
case study 231–2
seeds: EU uncertainty over GM seeds 140,
142; proposal to approve GM seeds in
the EU 129
stem cell research 164, 166; German
regulation 170–1; Swiss regulation 171
Switzerland: assisted reproductive
technologies (ART) regulation 171
systems biology 27–8

therapeutic cloning (TC) 166, 172, 173, 176
trade: China in commodities 120; effect
of China's GMO regulation on Canada
120, 121; implications of EU restrictions
on imports of GM products 131–2;
influence on response of countries to
GMOs 113–14; international safety
regulations inconsistencies 38; WTO
complaint against EU moratorium on
GM crops 95, 104–7, 131
trials: safety 20–2; *see also* field tests

United Kingdom: assisted reproductive
technologies (ART) regulation 171,
172–3; survey of biotechnology policy
actors 189–201
United States: abortion debate 174–5;
agricultural biotechnology 17; assisted
reproductive technologies (ART)
regulation 174–5; biotechnology
regulatory evolution 60–1, 63–4,
101–4; biotechnology regulatory
regime 58–9; Chinese imports 120,
121; Department of Agriculture
(USDA) 102; Environmental Protective
Agency (EPA) 102; federal regulation
of GMOs 103–4; Food and Drug
Administration (FDA) 102; GMO
policy compared with EU 95–107,
128; GMO production 101, 128;
survey of biotechnology policy actors
189–201; trade tensions with EU over
GM products 95, 104–5, 128; WTO
complaint against EU moratorium on
GM crops 95, 97, 104–7, 131

World Health Organization (WHO) 97
World Trade Organization (WTO):
complaint against EU moratorium on
GM crops 95, 97, 104–7, 131; EU
proposal to devolve regulatory authority
to member states 139–40; role 105